Matthias A. Fischer

# Surrogate-based uncertainty quantification and parameter optimization in simulations of the West African monsoon.

**Karlsruher Institut für Technologie**
**Schriftenreihe des Instituts für Technische Mechanik**

Band 41

Eine Übersicht aller bisher in dieser Schriftenreihe erschienenen Bände finden Sie am Ende des Buchs.

# Surrogate-based uncertainty quantification and parameter optimization in simulations of the West African monsoon.

by
Matthias A. Fischer

KIT Scientific Publishing

Karlsruher Institut für Technologie
KIT-Fakultät für Maschinenbau

Surrogate-based uncertainty quantification and parameter optimization in simulations of the West African monsoon

Zur Erlangung des akademischen Grades eines Doktors der Ingenieurwissenschaften von der KIT-Fakultät für Maschinenbau des Karlsruher Instituts für Technologie (KIT) genehmigte Dissertation

von Matthias A. Fischer, M.Sc.

Tag der mündlichen Prüfung: 3. Juni 2025
Hauptreferent: Prof. Dr.-Ing. Carsten Proppe
Korreferent: Prof. Dr. Peter Knippertz

**Impressum**

KIT Scientific Publishing

Karlsruher Institut für Technologie (KIT)
KIT Scientific Publishing
Straße am Forum 2
D-76131 Karlsruhe

KIT Scientific Publishing is a registered trademark of Karlsruhe Institute of Technology.
Reprint using the book cover is not allowed.

www.ksp.kit.edu

www.bibliothek.kit.edu/ksp.php | E-Mail: info@ksp.kit.edu | Shop: www.ksp.kit.edu



Print on Demand 2025 – Gedruckt auf FSC-zertifiziertem Papier

ISSN 1614-3914
ISBN 978-3-7315-1442-8
DOI 10.5445/KSP/1000182604

# Danksagung

Die vorliegende Arbeit entstand während meiner Tätigkeit als wissenschaftlicher Mitarbeiter am Institut für Technische Mechanik, Institutsteil Dynamik/Mechatronik, des Karlsruher Instituts für Technologie (KIT).

Mein besonderer Dank gilt Herrn Prof. Dr.-Ing. Carsten Proppe für die fachliche Betreuung und die Übernahme des Hauptreferats. Durch sein konstruktives Feedback, sein großes Vertrauen sowie die gewährten wissenschaftlichen Freiheiten konnte ich mich in einem inspirierenden Umfeld entfalten. Seine Offenheit ermöglichte es mir, im Rahmen der Themenfindung eine interdisziplinäre Zusammenarbeit mit dem Institut für Meteorologie und Klimaforschung zu initiieren.

Ich bedanke mich bei Herrn Prof. Dr. Peter Knippertz, mit dem mich seit unserer ersten Begegnung ein sehr vertrauensvolles Verhältnis verbindet. Ich schätze seine Bereitschaft, diese Zusammenarbeit einzugehen, die Übernahme des Korreferats sowie die kontinuierliche meteorologische Unterstützung, die er mir gewährt hat. Sein wertvoller Beitrag hat maßgeblich zum Erfolg dieser Kooperation beigetragen.

Mein besonderer Dank gilt auch Herrn Prof. Dr.-Ing. habil. Alexander Fidlin, der sich durch sein fachliches Engagement für die Forschungsthemen am Institut verdient gemacht hat und sich für eine positive Zukunft des Instituts, insbesondere im Dialog mit der Fakultät, einsetzt. Prof. Dr.-Ing. Wolfgang Seemann, der leider während meiner Promotionszeit verstorben ist, danke ich herzlich für sein großes Engagement im fachlichen Austausch und seinen vielfältigen Einsatz für das Institutsleben.

Ich möchte auch den emeritierten Professoren Prof. Dr.-Ing. Dr. h.c. Jens Wittenburg, Prof. Dr.-Ing. Dr. h.c. Jörg Wauer sowie Prof. Dr.-Ing. Walter Wedig, der leider wenige Wochen vor meiner Prüfung verstarb, danken, deren Teilnahme am Institutsleben und an den jährlichen Seminaren eine große Bereicherung darstellte.

Von meteorologischer Seite möchte ich insbesondere Dr. Gregor Pante, Dr. Roderick van der Linden und Dr. rer. nat. Alexander Lemburg danken, die mich bei der Durchführung der meteorologischen Modelläufe und der Interpretation der Ergebnisse intensiv unterstützt haben.

Meinem Bürokollegen Simon Keller danke ich für die sehr angenehme gemeinsame Zeit in den vergangenen Jahren und die gegenseitige Unterstützung bei fachlichen und organisatorischen Fragen.

Mein Dank gilt darüber hinaus allen Kolleginnen und Kollegen am Institut für Technische Mechanik, Institutsteil Dynamik/Mechatronik, sowie den Mitarbeitenden am Institut für Meteorologie und Klimaforschung Troposphärenforschung, die ich im Institutsleben und im Rahmen von Gruppenmeetings kennenlernen durfte. Der rege Austausch in Forschung, Lehre und auch im sozialen Miteinander hat meine Arbeit bereichert.

Ein besonderer Dank gilt meinem Bruder Christoph Fischer, dessen Weg sich trotz unterschiedlicher Studienrichtungen mit meinem kreuzte. Mit seinem informatikbezogenen Hintergrund und meteorologischen Fokus war er mir, vor allem in den ersten Jahren der Zusammenarbeit mit dem Institut für Meteorologie und Klimaforschung, eine wertvolle Unterstützung.

Ich bedanke mich zudem bei Prof. John Marsham von der School of Earth and Environment der University of Leeds sowie bei den Doktorandinnen und Doktoranden dieses Instituts, insbesondere Dr. Rachel Sansom und Erin Raif, die ich im Rahmen eines viermonatigen Forschungsaufenthalts kennenlernen durfte. Mit ihnen hatte ich einen wertvollen fachlichen und sozialen Austausch, der bis heute fortbesteht.

Karlsruhe, Juni 2025
Matthias A. Fischer

# Kurzfassung

Der westafrikanische Monsun (WAM) ist ein bedeutendes meteorologisches System mit weitreichenden Auswirkungen, dessen komplexe Wechselwirkungen Wetter- und Klimamodelle vor große Herausforderungen stellen. Unsicherheiten in Modellparametrisierungen, beispielsweise bei hochreichender Konvektion oder der Wolkenmikrophysik, beeinflussen die Vorhersagegenauigkeit erheblich. Diese Arbeit stellt einen Ansatz auf Basis von Ersatzmodellen vor, um diese Unsicherheiten zu quantifizieren und Modellparameter gezielt zu verbessern. Dazu wird ein Verfahren entwickelt, das den Parameterraum zunächst in einen gleichverteilten Eingangsparameterraum überführt, in dem Stichprobenverfahren sowie Surrogatmethoden angewandt werden. Ergebnisse aus Simulationen mit dem Icosahedral Non-Hydrostatic (ICON)-Modell des Deutschen Wetterdienstes werden mithilfe der Gaußprozess- und Hauptkomponentenregression analysiert, um den Rechenaufwand zu reduzieren. Der Einfluss der Modellparameter wird durch globale Sensitivitätsanalysen und Parameterstudien untersucht, gefolgt von Optimierungen unter Einbezug von meteorologischen Referenzdaten. Die Ergebnisse zeigen, dass unter den untersuchten Parametern insbesondere die Entrainment-Rate, die die Durchmischung zwischen aufsteigender Luft und Umgebungsluft beeinflusst, die Endfallgeschwindigkeit von Eispartikeln und der verdunstungsaktive Bodenanteil, der die Bodenfeuchte und damit die Verdunstung reguliert, den größten Einfluss auf das WAM-System haben. Eine Verringerung der Entrainment-Rate führt zu einer Verbesserung der Simulation von Niederschlag, bodennaher Luftfeuchte und Luftdruck. Allerdings sind Verbesserungen über alle Variablen hinweg begrenzt, was auf eine bereits gut abgestimmte Standardkonfiguration des ICON-Modells hindeutet. Die Ergebnisse verdeutlichen strukturelle Herausforderungen, etwa die begrenzte Möglichkeit, die gewünschten räumlichen Muster durch Parameteranpassungen zu erzielen, sowie Zielkonflikte zwischen verschiedenen meteorologischen Größen. Dies unterstreicht die Notwendigkeit, neben den Parameterwerten auch die Modellphysik und die räumliche Auflösung weiterzuentwickeln.

# Abstract

The West African monsoon (WAM) is a key climatic system with far-reaching impacts, and its complex interactions pose a challenge for weather and climate models. Uncertainties in model parameterizations, for example, in deep convection or cloud microphysics, significantly affect forecast accuracy. This dissertation presents a surrogate-based approach to quantify these uncertainties and systematically improve model parameters. To achieve this, a framework is developed that first transforms the parameter space into a uniformly distributed input space, where sampling techniques and surrogate methods are applied. Results from simulations with the Icosahedral Non-Hydrostatic (ICON) model of the German Meteorological Service are analyzed using Gaussian process regression and principal component regression to reduce computational costs. The influence of model parameters is examined through global sensitivity analyses and parameter studies, followed by optimizations incorporating meteorological reference data. The results show that among the investigated parameters, the entrainment rate, which governs the mixing between rising air and the surrounding atmosphere, the terminal fall velocity of ice particles and the evaporative soil surface fraction, which controls soil moisture and evaporation, have the greatest impact on the WAM system. Reducing the entrainment rate leads to a more accurate simulation of precipitation, near-surface humidity and mean sea-level pressure. However, improvements across all variables remain limited, suggesting that the default configuration of the ICON model is already well-calibrated. The findings highlight structural challenges, such as the limited ability to achieve desired spatial pattern changes through parameter adjustments and the trade-offs between different meteorological variables. This underscores the need to refine not only parameter values but also model physics and spatial resolution.

# Contents

# Appendix

# 1 Introduction

According to Moore's law [93], the number of transistors on an integrated circuit doubles approximately every two years. Despite this rapid growth in hardware capabilities, computational models across many scientific fields have not necessarily become faster; instead, they have grown more complex. For instance, while meteorological simulations in the 1990s used global grids with resolutions of hundreds of kilometers, today's models can achieve resolutions as fine as one kilometer [40].

The pursuit of greater accuracy in modeling complex systems requires not only substantial computational resources but also a more detailed representation of the underlying processes. Yet, certain processes – such as cloud formation – can never be fully captured with complete accuracy through fundamental physical equations and an atmospheric grid framework. As a result, these processes must be approximated using parameterizations, a common practice in meteorology that involves structural or empirical assumptions about unresolved phenomena. While aleatoric uncertainty arises from the inherent variability associated with the chaotic nature of the atmosphere, epistemic uncertainty is introduced through parameterizations, for example through the choice and specification of parameter values, reflecting limitations in our understanding of the underlying processes.

As models continue to grow in complexity, yet remain limited in their ability to fully represent all physical processes, there has been increasing interest in developing surrogate models. These simpler models can then be used to efficiently quantify uncertainties in more complex models and support their improvement, ultimately reducing those uncertainties. This work focuses on the West African monsoon (WAM), a prominent meteorological system known for its substantial prediction uncertainties [139]. The goal is to apply and advance state-of-the-art techniques, along with developing custom extensions, to assess model uncertainties and explore ways to enhance the accuracy of predictions for the system's key characteristics.

## 1.1 Motivation

The WAM is a major seasonal circulation system that brings extensive rainfall to West Africa during the boreal summer, typically peaking in August [48]. This precipitation is vital to the livelihoods of hundreds of millions of people, affecting agriculture, energy production, water resources, and public health, making it a key socioeconomic driver in the region [43, 107]. Despite its critical importance, simulations of the WAM across timescales from weather forecasts to climate predictions suffer from substantial uncertainties.

With respect to weather forecasts, ensemble rainfall predictions over tropical Africa exhibit the lowest skill of all tropical regions – a metric quantifying the accuracy of a forecast relative to a reference – and are often barely better than climatological forecasts [140], even after the application of statistical post-processing to reduce systematic errors. This poor performance can partly be attributed to uncertainties in the initial conditions, compounded by the sparse operational network in the region [23, 109]. Improving simulations over West Africa also requires enhancing numerical models themselves. Janicot, Lafore, and Thorncroft [54] argued that biases and uncertainties could be significantly reduced through a better understanding and more precise representation of processes occurring on weather timescales. Achieving such improvements, however, often requires extensive experimentation with model formulations and parameterizations, which can be computationally prohibitive.

To address this challenge, surrogate models have emerged over the past few decades as a valuable tool for exploring dependencies and uncertainties in complex computational systems. By providing a resource-efficient alternative to costly simulations, they enable detailed statistical investigations into how uncertain model parameters affect key system quantities [128]. These models are frequently used to optimize such model parameters; however, defining appropriate objectives, particularly in meteorology, remains a challenge due to the wide range of variables involved. As noted by Zängl [153], the process of tuning model parameters often involves subjective decisions, such as prioritizing specific forecast variables or regions, with the assignment of weights significantly influencing the resulting optimal parameters.

To date, no comprehensive study has explored uncertainty contributions from model parameters or examined parameter calibration specifically for the WAM. Moreover, previous meteorological research has not investigated how weight variation in multi-objective optimization (MOO) affects outcomes in this context. This work aims to address these gaps in current research.

## 1.2 State of Research

Accurate prediction of the WAM is vital due to its significant socioeconomic impacts, yet uncertainties in meteorological modeling remain, driven by the complexity of atmospheric processes. This section first introduces the meteorological system under study, providing an overview of the state of research and highlighting the key findings and challenges specific to this system. Building on this, the state of research in the methodological aspects – uncertainty quantification (UQ), surrogate modeling and parameter optimization – is presented, with a focus on their application to and relevance for the meteorological context, thereby establishing the motivation for advancing existing techniques.

The WAM, conceptually depicted in Fig. 1.1, constitutes a complex deep overturning circulation, with its formation, maintenance and variability governed by various regional and remote forcings [44]. One of its main initial drivers is the large temperature, and thus pressure gradient between the hot, dry and often dusty Sahara manifested in the Saharan heat low (SHL) and cooler, moister conditions over the tropical Gulf of Guinea. The marked discontinuity between these fundamentally different air masses, the Intertropical Discontinuity (ITD), which lies around 20°N during boreal summer, is associated with shallow and dry overturning [100, 132]. Abundant deep convection is rather observed in a band south of the ITD, often called monsoonal rain belt. There, mainly between 8°N and 13°N, the bulk of summertime precipitation is produced by frequently passing large convective systems with a high degree of organization [69, 70, 89].

The monsoonal rain belt is enclosed by two distinctive dynamical features, the African Easterly Jet (AEJ) to the north and the Tropical Easterly Jet (TEJ) to the south. The AEJ, a pronounced easterly jet at around 600–700 hPa maintained by the low-tropospheric meridional temperature gradient, regularly features wave disturbances. These so-called African Easterly Waves with wavelengths between 2000 km and 5000 km and periods of 2–7 days [8, 61, 115] strongly modulate convection, mainly by enhancing vertical wind shear to levels favorable for the generation of organized squall lines [22]. In the upper troposphere, the WAM circulation is characterized by a jet-like intensification of the tropical easterlies. This distinct easterly current observed between 5°N and 20°N, called TEJ, evolves over the South Asian monsoon system, where it is also strongest, and extends westward to Africa under gradual weakening [27]. Previous studies demonstrated that seasonal-mean WAM rainfall is strongly correlated with the intensity of the TEJ over West Africa [42]. At least on shorter timescales, the TEJ is, however, mainly thought of as a passive feature, which can intensify after periods

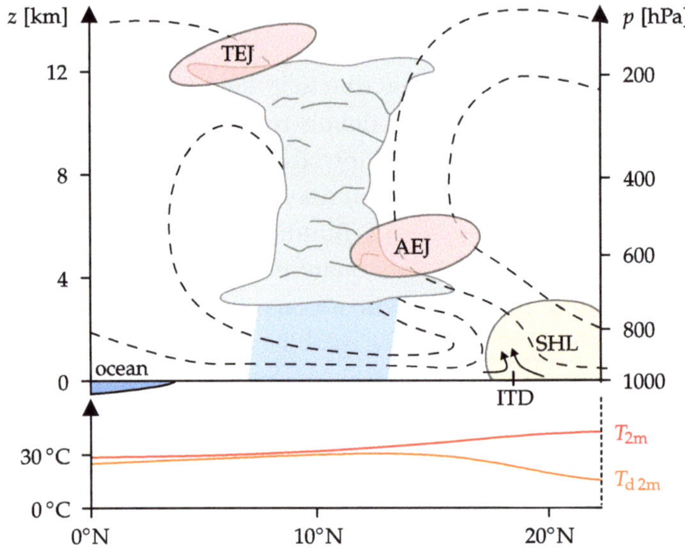


**Figure 1.1:** Schematic illustration of the WAM system in a height-latitude display (inspired by Fink et al. [24]) including the TEJ, the AEJ, the SHL, the ITD, 2 m temperature ($T_{2m}$) and 2 m dew point temperature ($T_{d\,2m}$). The main rainfall area is indicated by light blue shading. Circulation in the height-latitudinal plain is depicted through streamlines. The approximate latitudinal position of the Guinea Coast is also given.

of increased convective activity through the enhanced divergent outflow at upper levels [74].

Despite its outstanding importance for the region, simulations of the WAM spanning timescales from weather to climate are fraught with substantial uncertainties. For instance, the German Meteorological Service (DWD) currently uses a global grid with a mesh size of 13 km to achieve sufficiently accurate weather predictions. Even at these high resolutions, many atmospheric processes occur on a sub-grid scale and still need to be parameterized, since they cannot be explicitly resolved. Thus, such models involve a substantial number of parameters that must be carefully tuned. While some parameters represent quantities that are directly measurable or well-constrained by observations (e. g. the fall speed of ice), others cannot be directly observed and must be calibrated to ensure realistic model behavior at sub-grid scales (e. g. the soil moisture evaporation fraction). Vogel et al. [138, 139] showed that ensemble predictions of rainfall over tropical Africa have the lowest skill throughout the tropics and are often barely better than climatological forecasts [140], even after the removal of systematic errors through statistical post-processing. This poor performance is partly related to errors stemming from initial condition uncertainty in a region known for a sparse operational network [23, 109]. Moreover, there appear to be issues with data assimilation, as the availability of additional observations during field campaigns shows relatively small improvements [1, 75]. In weather forecasts, but also in mean-state focused simulations (beyond the problem of initial state uncertainty), the representation of the WAM and its features is affected by various model uncertainties. Shortcomings in adequately simulating small-scale diabatic processes, such as deep moist convection, not only directly affect rainfall prediction skill but also introduce errors in the overall WAM circulation. Like many tropical large-scale flows, the WAM is strongly driven by diabatic heating in the troposphere [83, 84]. Model-related uncertainties regarding the representation of deep convection and other physical processes are also reflected on climate timescales where many models struggle to realistically reproduce the rainfall distribution over the WAM region and its seasonal evolution [12, 135, 151]. Considerable problems are also evident on paleoclimate timescales with many models struggling to accurately describe the magnitude and time of precipitation changes of the African humid period during the Holocene, which amongst other things led to a Green Sahara [7, 11].

Reducing biases and uncertainties in model simulations over West Africa requires a more precise representation of processes on weather timescales. Janicot, Lafore, and Thorncroft [54] emphasized the need for detailed analyses at these shorter timescales to improve both weather forecasts and climate predictions. As mentioned in the paragraph above, especially the correct representation of diabatic processes, most of them indeed acting on short time and rather small spatial scales, still constitutes a major challenge. In this regard, a key problem are model uncertainties associated with grid resolution and parameter choices in the representation of sub-grid scale processes. For example, the explicit or parameterized representation of deep convection has a large effect on the amount, spatial distribution and diurnal cycle of precipitation, with substantial impacts on the large-scale dynamics and thermodynamics, even beyond the African continent [59, 83, 108]. Matsui et al. [90] found that the treatment of radiation in their model affects precipitation, low clouds and the entire WAM circulation, while Tchotchou and Kamga [131] highlighted the deficiencies of selected convection schemes to simulate the monsoon rainfall accurately. Flaounas, Bastin, and Janicot [25], Gbode et al. [32], and Klein et al. [62] considered microphysical, convective and boundary layer processes and found that variations in process parameters significantly influence the accuracy and spread of precipitation and other outputs. In other studies, effects of different meteorological phenomena and boundary conditions on the WAM were investigated. For instance, Kniffka, Knippertz, and Fink [63] highlighted that variations in low-level clouds can have a substantial impact on precipitation. Zheng and Eltahir [155] and Hopcroft et al. [51] investigated the influence of vegetation. The former examined variations in its meridional distribution on a weather timescale, while the latter identified a relationship between past vegetation coverage and mid-Holocene climate. Messager, Gallée, and Brasseur [92] found that sea surface temperature (SST) is a major driver of seasonal and interannual monsoon precipitation variability.

The aforementioned studies aimed to assess isolated relationships between specific physical schemes or their associated model parameters and simulated WAM quantities. A major limitation of this approach is the challenge of studying the combined effects of multiple sources of uncertainty simultaneously. Non-linear interactions and buffering effects will make it nearly impossible to deduce such effects from single-parameter perturbation experiments. Ideally, experiments would be conducted across a wide range of parameter combinations. However, this quickly becomes computationally expensive, as a sufficiently long simulation period is required to distinguish systematic differences from day-to-day weather variability.

To address these challenges, surrogate modeling techniques have gained increasing popularity across numerous scientific fields – such as engineering, meteorology, chemistry and economics – over the past decades. The terminology for surrogate models varies; in meteorology, they are often referred to as meta-models or emulators [10, 71, 85]. The relationship between a system, its model and the surrogate model is illustrated in Fig. 1.2. A system refers to the real-world process or phenomenon under investigation, governed by natural laws and mechanisms that are to be understood, simulated or predicted. A model is a mathematical or computational representation of such a system, incorporating the key variables, parameters and functional relationships believed to drive its behavior. In the present study, the system of interest is the atmosphere over West Africa, including the WAM, and the computational model employed is a limited-area numerical weather prediction model. Model parameters – often derived from observational data or heuristic assumptions – serve as inputs to the model and are inherently subject to uncertainty. These parametric uncertainties are typically of an epistemic nature, arising from incomplete knowledge of the system or model structure. In contrast, aleatoric uncertainty refers to the intrinsic variability of the system itself. Epistemic uncertainties are commonly represented by probability density functions (PDFs), causing model outputs to depend on the specific realizations of the uncertain input parameters. Since complex computational models are often too resource-intensive to support tasks requiring numerous model evaluations, surrogate models offer a practical alternative by providing efficient approximations of the original model. These surrogates are trained on a relatively small set of simulations, referred to as the experimental design [28, 122], and their associated outputs. Once constructed, surrogate models enable a wide range of applications, including parameter optimization [29] and UQ. Within this framework, UQ seeks to identify, characterize, and quantify

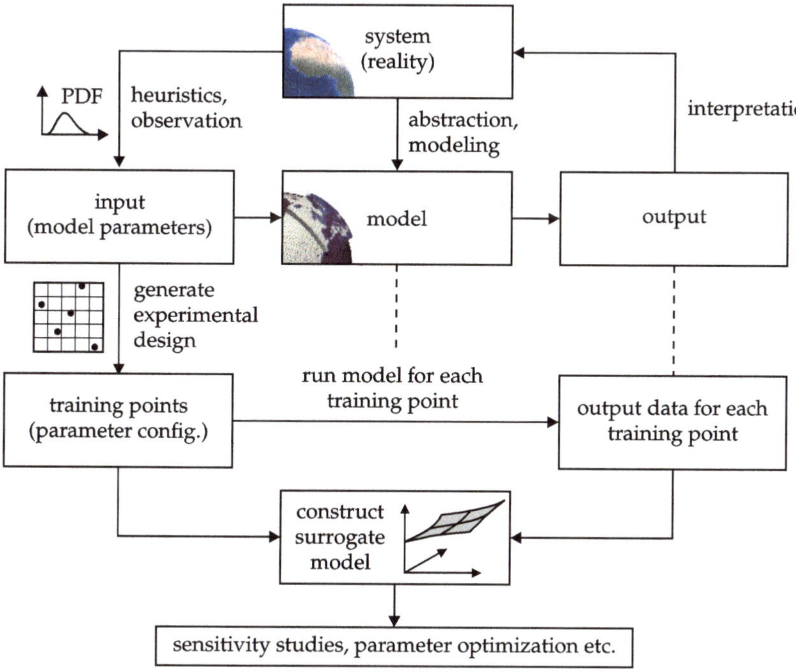


**Figure 1.2:** Schematic illustration of the relationship between a system, its model and the surrogate model.

uncertainties in model inputs and to assess their influence on model responses [67]. This framework supports a broad range of applications, including structural reliability analysis [73], global sensitivity analysis [120] and Bayesian methods for the calibration and validation of computational models [14].

Early surrogate techniques, such as polynomial regression, provide a straightforward approach to approximating system behavior. With the evolution of surrogate modeling, more advanced methods emerged, including kriging, which was introduced by French mathematician Georges Matheron in 1962 [86], building on earlier work by Danie G. Krige. Initially applied to estimate gold distributions in South Africa's Witwatersrand reef complex, kriging became a powerful tool for both approximation and UQ for the design and analysis of computer experiments [119]. This method, later known as Gaussian process regression (GPR) [113], has evolved into one of the most prominent approaches. Beyond approximating the output of a computational model, GPR also provides local estimates of its accuracy, enabling methodical extensions, such as in surrogate-based design optimization [96]. Other advanced techniques that have further expanded the range of surrogate modeling approaches include support vector regression [134], polynomial chaos expansions [148] and artificial neural networks [28].

The estimation of hyperparameters is central to GPR methods and is typically performed through likelihood maximization [113]. However, in high-dimensional input spaces or with large training datasets, this process becomes computationally expensive. To accelerate hyperparameter estimation, gradient-based methods are commonly employed. In the case of ordinary kriging, efficient expressions for computing optimal hyperparameters are well established and widely applied [113].

Other earlier methods, often not traditionally categorized as surrogate modeling techniques, can still effectively describe the response variables of a system. Statistical regression methods such as partial least squares regression [34] identify latent variables that maximize the covariance between predictors and responses, thereby addressing multicollinearity. Regularized regression techniques [156], including ridge regression, lasso regression and elastic net regression, improve model stability and perform variable selection. Dimensionality reduction methods, such as principal component analysis (PCA) [57] and factor analysis [46], transform or model variables to capture the most significant data structure. Principal component regression (PCR) [98] combines PCA with regression, traditionally in a linear framework.

Although the exact relationship between input and output quantities is not known in advance, it may be possible to assume certain underlying trends to improve prediction accuracy. These trends can be incorporated as basis functions by various surrogate modeling techniques, such as universal kriging [87]. Usually, simple basis functions (often low-order polynomials) are used to reduce complexity and avoid overfitting. However, in high-dimensional input spaces, higher-order multivariate polynomials can lead to a rapid increase in the number of terms, a challenge known as the curse of dimensionality. To address this, methods for polynomial basis selection, such as least-angle regression [19], can be applied. For example, Kersaudy et al. [60] proposed the LARS-Kriging-PC modeling method, where explicit multivariate basis functions for universal kriging are selected using least-angle regression based on polynomial chaos expansions. This method is designed to select the polynomials that contribute the most relevant information to the kriging model, effectively combining the strengths of GPR and polynomial chaos expansion. Similarly, blind kriging [58] identifies the underlying trend from data using Bayesian variable selection. Both methods require a predefined set of candidate basis functions, and the polynomials are chosen based on the regression technique rather than prior knowledge of the problem. However, as noted by Oakley [102], the choice of basis functions should, wherever possible, incorporate any prior beliefs or knowledge about the problem, such as the physical evolution of the output variable with respect to the input parameters.

In meteorological literature, GPR is the most widely used technique and has, for instance, been applied by Fletcher, Kravitz, and Badawy [26] to investigate the impact of aerosol forcing and atmospheric parameters on climate sensitivity, where two cloud- and convection-related parameters had the strongest influence. Similarly, Lee et al. [71] used GPR to study the cloud condensation nuclei (CCN) sensitivity to eight emission and microphysical process parameters and found that uncertainty in the sulphur emissions explains 80 % of the output variance. Radial basis functions were used by Müller et al. [97] to calibrate a model for methane emission-related parameters. Lu et al. [78] used advanced sparse grid interpolation as a surrogate model for the E3SM Land Model and employed quantum-behaved particle swarm optimization to identify optimal parameters. Alternative approaches include polynomial regression [3, 50, 99], polynomial chaos expansion [85] and neural networks [79]. Combinations of these methods have also been used. For instance, Ray et al. [114] applied Gaussian processes and polynomial regression for Bayesian calibration of hydrological parameters in the Community Land Model, while Chang et al. [9] combined PCA with GPR to develop surrogate models for calibrating the Greenland Ice Sheet model. Universal kriging has also been applied in several studies [17, 39, 144], using either linear or quadratic basis functions as trend functions.

When describing more than one output variable, multi-output modeling approaches have evolved to approximate all outputs with a single surrogate model, leveraging inter-output correlations to improve accuracy or sometimes reduce computational cost. For instance, Borchani et al. [5] summarized approaches such as multi-target regressor stacking, regressor chains and multi-output support vector regression. Co-kriging [88], an extension of GPR for multiple outputs, encompasses various methods for modeling inter-output dependencies, with the linear model of coregionalization being one prominent approach. Additionally, PCA [35] and PCR [36] have been adapted for spatial data, gaining popularity in image processing applications where principal components are employed in multivariate regression. With advancements in computational power and data availability, neural network-based approaches have become more prominent for modeling relationships between multiple inputs and outputs [149].

Additionally, parameter optimization studies have been extended to multiple outputs leading to MOO problems [29]. Particularly in meteorology, defining suitable objectives remains challenging due to the diverse range of variables involved. When optimizing for multiple objectives simultaneously, compromises can be identified via Pareto fronts. On such Pareto fronts, an improvement in one objective necessitates the deterioration of at least one other objective. Surrogate-based optimization has proven particularly effective for these complex MOO problems, which often require high computational effort.

Several studies have focused on identifying Pareto fronts using surrogate models [41, 124]. Other studies have simplified the analysis by predefining the weighting of objectives, thereby reducing the problem to a single-objective optimization problem. For example, Cinquegrana et al. [10] assigned equal weights to all objectives, although the weights may have a substantial impact on the optimization results.

Highlighting the complexity of improving the Icosahedral Non-Hydrostatic (ICON) model – the operational forecast system of the DWD – by tuning its parameters, Zängl [153] noted that many parameters have a range of values with no clear optimum. Modifying a parameter may improve model performance in some regions or seasons while degrading it in others, or may enhance certain forecast variables at the expense of others. Traditionally, parameter tuning has been conducted by experts without a unified framework. History matching [145] has emerged as a prominent technique for quantifying parameter uncertainties by systematically exploring a range of plausible model configurations and eliminating parameter sets that fail to reproduce observations within acceptable tolerances. This approach is particularly useful in the tuning of climate model parameters, such as those governing aerosol-cloud interactions, where direct measurements are often challenging to obtain [72]. Over the past decades, automatic calibration techniques have emerged, incorporating data assimilation methods into operational weather forecasts. A comprehensive review by Ruiz, Pulido, and Miyoshi [118] focused on these techniques and Zängl [153] described the current implementation in the ICON model. Ruiz, Pulido, and Miyoshi [118] emphasized that objective optimization often becomes infeasible when complex numerical models and a large number of parameters are involved. Nevertheless, when multiple forecast variables are available, the selection of objectives for optimization studies remains a critical task. Instead of combining meteorological variables independently in the MOO process, different criteria might be employed, such as the energy norm used by Ollinaho et al. [104]. However, depending on the amount and type of available data, such sophisticated criteria are rarely possible in meteorological surrogate-based optimization studies. As Zängl [153] argued, trying to select optimal parameters inherently involves subjective decisions, e. g. prioritizing certain forecast variables or regions. Furthermore, there is a risk that calibrated model processes might compensate for model errors originating from different parameterizations. Therefore, the investigation of the impact of such decisions, such as different objective weights, remains a crucial task and an open topic of research.

Surrogate model construction relies on appropriate sampling strategies to define training points, i. e. model parameter combinations. Classical sampling techniques, such as Monte Carlo (MC) simulation and full factorial sampling, are often impractical due to the curse of dimensionality, high computational cost, clustering and uneven coverage. Various alternative approaches have been developed to enhance specific criteria for training points, particularly space-filling designs, which aim to optimally cover the input space [110]. Later advancements also considered improvements related to model evaluations. In meteorological literature, Latin hypercube sampling (LHS) [95] has been frequently used [71, 79], including extensions such as maximin- or minimax-distance criteria and hierarchical sample refinements. In addition, Williamson [146] applied k-extended Latin hypercubes for GPR. Some studies applied alternative methods, such as Quasi-MC sampling [114] and polynomial chaos-based approaches [85]. Furthermore, sequential algorithms can supplement the base design by adding training points in regions where higher model accuracy is needed. When applying space-filling designs, training points in the input space are typically generated within predefined bounds. However, defining these bounds can be challenging. If they are too wide, fewer training points are placed in high-probability regions, an issue that becomes particularly severe in high-dimensional input spaces. Conversely, if the bounds are too narrow, the training points fail to cover the tails of the input distributions, which may still be of interest.

Consequently, a uniform distribution of training points in the input space is not always desirable [80]. For example, in reliability engineering, specific input domains such as the failure region are of primary interest [73], often requiring techniques like importance sampling [82]. Similarly, when models struggle to accurately capture the response surface in certain regions, additional training points may need to be added sequentially [110]. Another method to sample from a desired probability distribution is inverse transform sampling [16]. The idea of generating more samples in regions associated with higher probabilities using inverse transform sampling has often been used in the context of UQ and surrogate modelling to make models in such regions more accurate [2, 154]. For GPR, V. Roshan Joseph and Myers [133] introduced enhanced designs that, in addition to generating samples based on the PDFs of uncertain parameters – referred to as noise factors – also incorporate the surrogate prediction variance for training point selection. Thus, using a hybrid design, a compromise is found between a high training point density in regions with higher probabilities and the avoidance of high prediction variances in regions with lower training point density. More recently, Quinlan, Movva, and Perfect [111] proposed an active learning setup for GPR considering surrogate prediction variance combined with a weight function. In the latter two works, a stationary kernel function was assumed in the whole input space. However, Gaussian processes with stationary kernel functions may struggle to accurately capture the model behaviour, particularly if there are very few training points in the distribution tails [133]. It thus remains an open research question on how non-stationary kernel functions could handle this issue.

To facilitate surrogate modeling, it is well established to normalize or standardize data [55]. This improves numerical stability and ensures well-posedness when applying surrogate models. For this purpose, it is appropriate to both scale inputs and outputs of a model, particularly if values vary significantly in magnitude. Surrogate models, such as GPR with isotropic kernel functions, may otherwise struggle to account for such disparities, which can negatively affect their performance. Appropriate scaling can also accelerate model training, such as hyperparameter optimization for GPR. Box and Tidwell [6] and Viana, Gogu, and Goel [136] suggested to also use transformations based on simple mathematical relationships in the inputs such as low order polynomials to improve model performance. The idea of using such transformations is related to the motivation of using a feature space for GPR. Here, the inputs are projected into a high dimensional space using a set of basis functions and then apply the linear model in this space instead of directly on the inputs themselves. Using the *kernel trick*, products between functions in the feature space can then be replaced by a function which only depends on a distance measure between two input points [113].

In GPR, stationarity is commonly assumed, often as a reasonable initial approach when little is known about the underlying physical problem. However, if this assumption does not hold, an input space transformation may exist that maps the problem to a new space where stationarity can be assumed. The concept of transforming input spaces to achieve stationarity and isotropy in a new domain was introduced by Sampson and Guttorp [121], who applied thin-plate splines to realize such a mapping. Schmidt and O'Hagan [123] extended this method by employing a Bayesian approach, where the mapping is described by a function with a Gaussian process prior. Both methods were developed for two-dimensional input spaces within the context of geostatistics. Generally, such transformations can negatively affect prediction accuracy of surrogate models which is rarely investigated in detail. Similarly to transforming the input space to a new space where stationarity is assumed to hold, one could consider transforming it to ensure the suitability of a desirable space-filling design, such as a uniform density of training points.

For GPR, an alternative to transforming the input space to assume stationarity is to use non-stationary kernel functions in the original input space. Gibbs [37] proposed an approach for incorporating non-stationary kernels, with a simplified version demonstrated by Xiong et al. [147]. Their method constructs a density function to describe variations in smoothness with respect to input parameters. This density function is then used to define a nonlinear mapping to an input space where uniform smoothness – and thus a stationary kernel – can be assumed. Paciorek and Schervish [106] proposed explicitly modeling input-dependent smoothness using an additional GPR model.

Various software packages have also been developed to facilitate specific surrogate-based optimization procedures, such as the toolkit by Watson-Parris et al. [142], which supports model calibration using a range of surrogate models (e. g. GPR), and the parameterization improvement tool by Couvreux et al. [13], which incorporates GPR, history matching and other techniques. However, such software packages are usually rather limited in their flexibility and are not able to adress the methodical advancements proposed in this work, e. g. the possibility of defining ansatz functions for universal kriging, variations in objective weights in MOO etc.

No previous studies have comprehensively investigated uncertainty contributions and parameter optimization for the WAM system. Methodologically, past research had to balance generating samples according to parameter PDFs with the challenges surrogate models face in accurately capturing regions with low training point coverage or when using transformed parameter spaces. Furthermore, no studies have examined variations in objective weights in MOO in such detail.

## 1.3 Objectives and Structure

Following the motivation and summary of the current state of research, this section outlines the structure and objectives of the work. The objectives are defined based on the following open research needs:

- Development of surrogate models as cost-effective alternatives for representing characteristics of the WAM system, utilizing the ICON model.
- Consideration of both reduced scalar quantities of interest (QoIs) and meteorological fields as target outputs for the surrogate models.
- Advancement of the surrogate modeling process by establishing a comprehensive framework for the considered meteorological model that leverages the benefits of traditional experimental designs while ensuring a high predictive accuracy.
- Extension of the equations for gradient-based hyperparameter estimation from ordinary kriging to universal kriging.
- Execution of MOO in meteorological studies, including variations in the weights assigned to the individual objectives.

In this work, the foundational methods are first introduced and subsequently applied to the meteorological system. The results are then thoroughly interpreted and analyzed within the context of meteorological science.

The structure of this work is as follows. Chapter 2 establishes the foundations for the study, covering fundamental topics in probability theory, GPR, PCR, validation methods for surrogate models and global sensitivity analysis (GSA). Chapter 3 discusses advanced concepts in surrogate modeling and optimization. In Section 3.1, the transformation of the input parameter space, conducted with respect to probability density functions, is presented along with its implications for surrogate models, where for universal kriging, the equations for gradient-based hyperparameter estimation are derived. In Section 3.2, MOO concepts with varying objective weights are proposed.

The methods presented are applied to simulations of the WAM in Chapter 4. Section 4.1 describes the ICON model, while Section 4.2 introduces the meteorological data under consideration. In Section 4.3, surrogate models for the system are developed, utilizing GPR for scalar QoIs and PCR for meteorological fields. In Section 4.4, the surrogate models are used to examine the influence of model parameters, through both GSA and parameter studies, where individual parameters are varied separately to observe their effects on QoIs and meteorological fields. Section 4.5 uses the surrogate models to conduct optimization studies, with the results analyzed and discussed. Chapter 5 provides the conclusion of this work, including a summary and further perspectives.

# 2 Fundamentals

This chapter presents the mathematical foundations underlying the meteorological analysis conducted in this work. First, the relevant fundamentals of probability theory are introduced. Next, GPR and PCR are presented as surrogate modeling approaches for scalar QoIs and field data, respectively. The following section details the validation techniques employed to assess the performance and reliability of these surrogate modeling methods. Furthermore, the fundamentals of GSA are outlined, enabling the quantification of parameter influence on scalar outputs.

## 2.1 Fundamental concepts of probability theory

The uncertainty measures considered in this study originate from various sources. To provide a structured framework, these sources are first classified following common approaches in the literature. Subsequently, probability distributions and the probability integral transformation (PIT) are introduced as fundamental concepts. Finally, different sampling techniques are presented to illustrate their role in surrogate modeling and UQ.

### 2.1.1 Sources of uncertainty

Uncertainty is an inherent aspect of many real-world phenomena and engineering systems. Understanding and quantifying uncertainty is crucial for making reliable predictions, informed decisions, and developing robust models. Uncertainties can be broadly categorized into two types: epistemic and aleatoric [129].

Epistemic uncertainty – from the Greek επιστήμη, meaning knowledge – arises from a lack of knowledge or information about the system or process under study. It is often referred to as reducible uncertainty because it can be decreased with additional data, improved models, or more accurate measurements. Epistemic uncertainty is typically associated with factors such as incomplete understanding of underlying mechanisms, insufficient data, or simplifications made in modeling complex systems. It is often subdivided into *model form* and *parametric* uncertainty, where the model form uncertainty relates to structurally imprecise models and parametric uncertainty assumes a correct model form but imprecise parameter values in the model. In meteorological models, parametric uncertainties can be attributed to model parameters that affect physical processes, such as the terminal fall velocity of ice crystals. This quantity cannot be represented by the fundamental physical model equations on the discrete model grid with its limited resolution and is therefore defined in physical schemes. It depends on various factors, such as the particle size and shape, or the constitution and behavior of the air. The more precisely these circumstances can be estimated or the better the fall velocity can be measured, the more these epistemic uncertainties can be reduced.

Aleatoric uncertainty – from the Latin *alea*, meaning a die –, on the other hand, stems from inherent variability or randomness in the system or process. This type of uncertainty is also known as irreducible uncertainty because it cannot be eliminated, even with perfect information. Aleatoric uncertainty is often encountered in natural phenomena and processes where stochasticity plays a significant role, such as the rolling of dice, weather patterns, or the inherent variability in material properties. In weather systems, this type of uncertainty can be attributed to the chaotic nature of the system. Even if parameters associated with epistemic uncertainties were precisely known, aleatoric uncertainties in measured variables such as temperature or precipitation would still be present, typically increasing with time from an initial state of the system.

### 2.1.2 Probability distributions

Probability distributions are mathematical functions that describe the likelihood of occurrence of different values or outcomes, whether these arise from random variables, simulations, or real-world processes. Formally, let $(\Omega, \mathcal{F}, P)$ be a probability space, where $\Omega$ represents the sample space (the set of all possible outcomes), $\mathcal{F}$ denotes a $\sigma$-algebra of events (subsets of $\Omega$), and $P : \mathcal{F} \rightarrow [0, 1]$ is the probability measure assigning probabilities to events. A random variable $X$ is defined as a measurable function $X : \Omega \rightarrow \mathbb{R}$. Random variables can be classified as discrete, with a countable set of possible outcomes, or continuous, with an uncountable set of possible outcomes. The support of $X$, denoted $D_X$, is the smallest closed subset of $\mathbb{R}$ such that $P(X \in D_X) = 1$. In practice, for discrete random variables, the support is often taken as the set of values with positive probability, i. e. $D_X = \{x \in \mathbb{R} : P(X = x) > 0\}$.

For continuous random variables, the probability of $X$ assuming a specific value is zero. Instead, its behavior is described by the PDF, $f_X(x)$, which satisfies the properties

$$f_X(x) \geq 0 \quad \text{for all } x \in \mathbb{R} \quad \text{and} \quad \int_{-\infty}^{\infty} f_X(x)\,\mathrm{d}x = 1\,.$$

The cumulative distribution function (CDF), $F_X(x)$, gives the probability that $X$ takes a value less than or equal to $x$. For continuous random variables, it is defined as

$$F_X(x) = P(X \leq x) = \int_{-\infty}^{x} f_X(t)\,\mathrm{d}t\,,$$

and satisfies the properties $F_X(-\infty) = 0$, $F_X(\infty) = 1$, and it is a non-decreasing function. The support of a continuous random variable is commonly defined as $D_X = \{x \in \mathbb{R} : f_X(x) > 0\}$, identifying the region where the density function is strictly positive. The inverse of the CDF, known as the quantile function or percent-point function (PPF), provides the value $x$ corresponding to a given probability level $u$. That is, given $u = F_X(x)$, the inverse relationship is expressed as $x = F_X^{-1}(u) = Q(u)$.

In multivariate cases, the joint distribution of two or more random variables describes their combined behavior. For two random variables $X$ and $Y$, the joint PDF $f_{X,Y}(x,y)$ satisfies $f_{X,Y}(x,y) \geq 0 \quad \text{for all } (x,y) \in \mathbb{R}^2$ and $\int_{-\infty}^{\infty}\int_{-\infty}^{\infty} f_{X,Y}(x,y)\,dx\,dy = 1$. The joint CDF $F_{X,Y}(x,y)$ provides the probability that $X \leq x$ and $Y \leq y$:

$$F_{X,Y}(x,y) = P(X \leq x, Y \leq y) = \int_{-\infty}^{x}\int_{-\infty}^{y} f_{X,Y}(t,u)\,\mathrm{d}t\,\mathrm{d}u\,.$$

The marginal distributions of $X$ and $Y$ are obtained by integrating out the other variable:

$$f_X(x) = \int_{-\infty}^{\infty} f_{X,Y}(x,y)\,\mathrm{d}y\,, \quad f_Y(y) = \int_{-\infty}^{\infty} f_{X,Y}(x,y)\,\mathrm{d}x\,.$$

Conditional probability is the probability of an event occurring given that another event has already occurred. For random variables $X$ and $Y$, the conditional probability distribution of $X$ given $Y = y$ is denoted by $f_{X|Y}(x|y)$ and is defined as

$$f_{X|Y}(x|y) = \frac{f_{X,Y}(x,y)}{f_Y(y)} \quad \text{for } f_Y(y) > 0\,.$$

Statistical measures such as expectation and variance provide insights into the distribution's central tendency and spread. For a continuous random variable $X$, the expectation or mean is

$$\mathbb{E}[X] = \int_{-\infty}^{\infty} x f_X(x)\,\mathrm{d}x\,,$$

while the variance is

$$\mathrm{Var}(X) = \mathbb{E}[(X - \mathbb{E}[X])^2] = \mathbb{E}[X^2] - (\mathbb{E}[X])^2\,.$$

The standard deviation, $\sigma_X = \sqrt{\mathrm{Var}(X)}$, quantifies the average deviation of $X$ from its mean.
The covariance between two random variables $X$ and $Y$, defined as

$$\mathrm{Cov}(X,Y) = \mathbb{E}[(X - \mathbb{E}[X])(Y - \mathbb{E}[Y])] = \mathbb{E}[XY] - \mathbb{E}[X]\mathbb{E}[Y]\,,$$

measures their linear dependency. A positive covariance indicates that the variables tend to increase together, while a negative covariance implies an inverse relationship.

### 2.1.3 Probability integral transformation

The PIT enables the mapping of random variables to a uniform distribution. If $X$ is a random variable with a continuous CDF $F_X(x)$, then the transformed random variable $U = F_X(X)$ follows a uniform distribution on the interval $[0,1]$, i. e. $U \sim \mathrm{Uniform}(0,1)$. The isoprobabilistic transformation generalizes this concept, enabling the mapping of random variables from one probability distribution to another. In the univariate case, this transformation involves applying the CDF of the original variable followed by the PPF of the target distribution: $Y = F_Y^{-1}(F_X(X))$, where $F_X$ is the CDF of the original random variable $X$, and $F_Y^{-1}$ is the PPF of the target random variable $Y$.

In the multivariate case with dependent variables, the Rosenblatt transformation [117] provides a method for transforming the physical input vector $\mathbf{x} = (x_1, x_2, \ldots, x_p)^\top$ into an independent and identically distributed (i. i. d.) uniform input vector $\mathbf{u} = (u_1, u_2, \ldots, u_p)^\top$, where the input space dimension is $p$. Let $\mathcal{X} = (X_1, X_2, \ldots, X_p)^\top$ be a random vector in the physical input space with joint CDF $F_{\mathcal{X}}(\mathbf{x})$.

The Rosenblatt transformation $\mathbf{u} = \mathcal{T}_{\text{ros}}(\mathbf{x})$ is defined as:

$$
\begin{aligned}
u_1 &= P(X_1 \leq x_1) = F_{X_1}(x_1)\,,\\
u_2 &= P(X_2 \leq x_2 \mid X_1 = x_1) = F_{X_2|X_1}(x_2 \mid x_1)\,,\\
&\;\vdots\\
u_p &= P(X_p \leq x_p \mid X_{p-1} = x_{p-1}, \ldots, X_1 = x_1) = F_{X_p|X_{p-1},\ldots,X_1}(x_p \mid x_{p-1}, \ldots, x_1)\,,
\end{aligned}
$$

where $F$ denotes the respective (conditional) CDF. It can be shown that the transformed random vector $\mathcal{U} = (U_1, U_2, \ldots, U_p)^\top = \mathcal{T}_{\text{ros}}(\mathcal{X})$ is i. i. d. and uniformly distributed over the $p$-dimensional unit hypercube. The CDF functions $F$ can be derived from the joint PDF $f_{\mathcal{X}}(\mathbf{x})$ [91]. It should be noted that the Rosenblatt transformation is not unique, as it depends on the ordering of the input variables $X_i$. This ordering determines the sequence of conditional distributions and thus influences the resulting transformation.

### 2.1.4 Sampling techniques

Sampling techniques are methods designed to select a representative subset of the input space for analysis, aiming to capture the system's essential characteristics while minimizing computational or experimental costs. In this work, these techniques are employed to generate a set of training points – referred to as an experimental design or design of experiments – for simulations of a computational model. The simulation results, paired with their corresponding training points, serve as the foundation for constructing surrogate models.

Let $\mathbf{u}_i = \{u_{ij}, j = 1, \ldots, p\} \in \mathbb{R}^p$ denote a training point, i. e. a vector of one input parameter configuration (see Fig. 1.2), in the $p$-dimensional i. i. d. uniform input space of the domain $\mathcal{D} = [0, 1]^p$. The collection of sampling points determined by the sampling technique is denoted by $\mathbf{U} = \{\mathbf{u}_i, i = 1, \ldots, n\}$.

Sampling techniques are often designed to achieve desirable properties such as good coverage or randomness. When the system is treated as a black box, sampling methods that focus on generating space-filling designs are particularly relevant. The quality of space-filling designs can be evaluated using different approaches, broadly categorized into uniformity-based and distance-based criteria [31]. Uniformity-based criteria assess the discrepancy, which measures how far a given design deviates from an ideal uniform design. In a uniform design, the number of sample points within a subspace $\Delta\mathcal{D}$ of a domain $\mathcal{D}$ is proportional to the hyper-volume $V(\Delta\mathcal{D})$. Notable examples for such measures include the star discrepancy and the $L_2$-discrepancy. On the other hand, distance-based criteria evaluate space-filling designs by analyzing the distances between pairs of sample points. For instance, the potential energy (PE) criterion

$$\mathrm{PE}(\mathbf{U}) = \sum_{i=1}^{n} \sum_{j=i+1}^{n} \frac{1}{d(\mathbf{u}_i, \mathbf{u}_j)^2}$$

uses the Euclidean distance $d(.,.)$ between pairs of points to evaluate the design. By minimizing the PE, an evenly distributed design is achieved, as the criterion discourages clustering of points. Other notable distance-based criteria include the maximin and minimax criteria [56], defined as follows:

$$\mathrm{maximin}(\mathbf{U}) = \max_{\mathbf{U}} \left( \min_{i \neq j} d(\mathbf{u}_i, \mathbf{u}_j) \right) ,$$

$$\mathrm{minimax}(\mathbf{U}) = \min_{\mathbf{U}} \left( \max_{\mathbf{u} \in \mathcal{D}} d(\mathbf{u}, \mathbf{U}) \right) .$$

The maximin criterion aims to maximize the minimum Euclidean distance between any two sample points, ensuring a well-spread design and avoiding clustering. In contrast, the minimax criterion seeks to minimize the maximum distance from any point in the input space to its nearest sample $d(\mathbf{u}, \mathbf{U})$, thereby reducing sparse regions in the design. Other distance-based criteria include the minimum spanning tree criterion, the point-to-point distance criterion, and others, each offering different perspectives on assessing the quality of space-filling designs.

A selection of prominent sampling methods is illustrated in Fig. 2.1. A classical example of a space-filling criterion-based sampling method is grid sampling, where the domain is discretized into a grid, and points are sampled at grid intersections (Fig. 2.1a). This approach becomes computationally expensive in high dimensions due to the curse of dimensionality. Additionally, projecting such training designs along individual parameters collapses all training points along that dimension into single points. This can significantly reduce the information content if a parameter has only a marginal effect on the system's behavior.

Monte Carlo (MC) methods (Fig. 2.1b) generate sample points using pseudo-random number generators. The randomness aims to approximate a uniform distribution across the domain, but issues such as clustering or underrepresentation can arise. Stratified MC sampling (Fig. 2.1c) addresses these issues by dividing the domain $\mathcal{D}$ into multiple strata and applying MC sampling within each stratum, effectively reducing variance and improving uniformity.

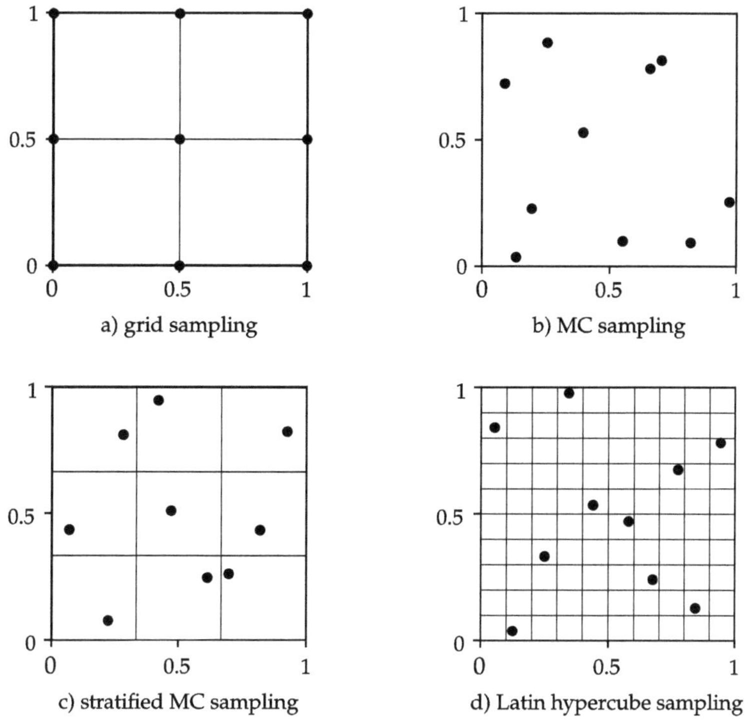


**Figure 2.1:** Realizations of different sampling techniques in a 2-dimensional $[0,1]^2$ space.

As one prominent and widely used special case of stratified MC sampling, Latin hypercube sampling (LHS), introduced by [81], effectively addresses the limitations of the aforementioned methods. In a $p$-dimensional design space $\mathcal{D}^p$, each dimension is divided into $n$ equal bins of edge length $\frac{1}{n}$, creating $n^p$ hypercubes. Subsequently, $n$ sample points are arranged as an $n \times p$ matrix $\mathbf{L} = [\mathbf{u}_1, \mathbf{u}_2, \ldots, \mathbf{u}_n]^\top$, where each column represents an input dimension and each row a sample point. Then, $\mathbf{L}$ is a $p$-dimensional Latin hypercube design (LHD) of size $n$ if no two elements in any column fall into the same bin. For example, in a two-dimensional design space ($p = 2$), the space can be visualized as a chessboard with $n \times n$ fields, where samples are placed such that each row and column of the board contains exactly one sample point (Fig. 2.1d). Since the configuration and sample placement within bins remain random, LHS inherently combines both random and space-filling properties. Extensions such as minimax LHS or maximin LHS improve space-filling properties by selecting the design that optimizes the respective criterion from a set of candidate LHDs.

Other space-filling methods include quasi-MC sampling, which employs deterministic, low-discrepancy sequences such as Halton, Hammersley and Sobol sequences. These sequences inherently aim for space-filling properties but do not formally quantify them during the sampling process. Consequently, their effectiveness depends on the problem's dimensionality, the number of samples, and their sensitivity to aliasing, which can introduce computational inaccuracies when constructing surrogate models.

More advanced methods incorporate system information to improve specific criteria. Adaptive sampling methods iteratively refine the sample set based on insights gained from previous samples, making them particularly valuable for surrogate modeling and optimization, where initial sample size estimation is challenging, and simulation outcomes can guide the process [157].

## 2.2 Gaussian process regression

Originating from geostatistics, GPR has proven to be a flexible, powerful, and accurate technique that is widely applied across various scientific disciplines, including meteorology. The formulation of GPR can be approached from two distinct perspectives: the weight-space view and the function-space view, both of which lead to the same regression equations [113]. In the weight-space view, GPR is derived from a linear regression model with a Bayesian treatment of the weights, where the uncertainty over weights is explicitly incorporated by averaging over all possible weight configurations. In contrast, the function-space view directly models functions as Gaussian processes, providing not only the predictive mean but also the predictive variance. This variance quantifies the uncertainty associated with predictions, which is a key advantage of Gaussian processes in regression tasks. Given the advantages of the function-space formulation, the equations presented in the following sections are derived from this perspective.

The aim of GPR is to build a surrogate model $\hat{y}$ for a scalar model output, i. e. a scalar QoI, based on the training data $\{\mathbf{U}, \mathbf{y}\}$. For numerical stability, the data $\mathbf{y}$ is assumed to be standardized or normalized, which must be ensured beforehand [55]. The objective is to determine the predictive mean and predictive variance at a set of input points, referred to as prediction points $\mathbf{U}_\star = \{\mathbf{u}_{\star\,i},\ i = 1 \ldots l\}$.

### 2.2.1 Simple kriging

The prior for the spatial process is a zero mean Gaussian process

$$\hat{y}(\mathbf{u}) \sim \mathcal{GP}(0, k(\mathbf{u}, \mathbf{u}')) \,.$$

with covariance function $k(\mathbf{u}, \mathbf{u}')$, also known as kernel function. The kernel function describes the dependence structure between values of the stochastic process at different points, usually depending on their distance.
Here, the anisotropic form of the radial-basis function

$$k(\mathbf{u}, \mathbf{u}') = \sigma_1 \exp\left(-\sum_{i=1}^{p}\left(\frac{|u_i - u_i'|}{\ell_i}\right)^2\right) + \sigma_2^2 \delta(\mathbf{u} - \mathbf{u}') \tag{2.1}$$

with respect to hyperparameters [1] $\boldsymbol{\theta} = \{\ell_i, i = 1 \dots p; \sigma_j, j = 1, 2\}$ is used. The Kronecker delta function $\delta(\mathbf{u} - \mathbf{u}')$ becomes 1 if $\mathbf{u} = \mathbf{u}'$ and 0 otherwise. The parameters $\ell_i$ can be interpreted as length scale parameters to allow for different smoothness between input dimensions. The parameter $\sigma_1$ represents the signal variance, which scales the output of the kernel function and controls the amplitude of the function variations and $\sigma_2$ represents the noise variance in the observations where the noise is assumed to be i. i. d. with a normal distribution and thus being refered to as white noise. This noise parameter can account for measurement errors or for aleatoric uncertainties in the simulations of a computational model.

The covariance matrices, defined by kernel function evaluations at the training points $\mathbf{U}$ and prediction points $\mathbf{U}_\star$, are given by

$$\begin{aligned}
\mathbf{K} &= \{K_{ij} &&= k(\mathbf{u}_i, \mathbf{u}_j), && i = 1 \dots n, && j = 1 \dots n\}\,, \\
\mathbf{K}_\star &= \{K_{\star,ij} &&= k(\mathbf{u}_i\,, \mathbf{u}_{\star\, j}), && i = 1 \dots n, && j = 1 \dots l\}\,, \\
\mathbf{K}_{\star\star} &= \{K_{\star\star,ij} &&= k(\mathbf{u}_{\star\, i}, \mathbf{u}_{\star\, j}), && i = 1 \dots l, && j = 1 \dots l\}\,.
\end{aligned}$$

The joint distribution of training data $\mathbf{y}$ and prediction data $\mathbf{y}_\star$ is then given by

$$\begin{pmatrix}\mathbf{y}\\ \mathbf{y}_*\end{pmatrix} \sim \mathcal{N}\left(\begin{pmatrix}0\\0\end{pmatrix}, \begin{pmatrix}\mathbf{K} & \mathbf{K}_\star\\ \mathbf{K}_\star^\top & \mathbf{K}_{\star\star}\end{pmatrix}\right) . \tag{2.2}$$

[1] The denotation *hyperparameters* is used in this context for variables that describe higher-level settings that govern the behavior and performance of the model, whereas *parameters* typically refer to the coefficients within the model, such as weights in a neural network or coefficients in a linear model.

Note, that the covariance matrix of training points $\mathbf{K}$ is sometimes replaced by $\mathbf{K} + \alpha\mathbf{I}$ to ensure positive definiteness, allowing for a numerically stable inversion. In this work, the error term is instead incorporated as white noise in the kernel function using $\sigma_2$ (Eq. 2.1). This approach provides greater flexibility and a clearer separation between signal and noise while explicitly accounting for observational noise or, in computational models, aleatoric uncertainties.

The posterior distribution is obtained by conditioning the joint Gaussian prior distribution on the observations as

$$\mathbf{y}_* \mid \mathbf{U}_\star, \mathbf{U}, \mathbf{y} \sim \mathcal{N}(\mathbf{K}_\star^\top \mathbf{K}^{-1}\mathbf{y}, \mathbf{K}_{\star\star} - \mathbf{K}_\star^\top \mathbf{K}^{-1}\mathbf{K}_\star)\,.$$

From this, the predictive mean (regression model) and predictive covariance matrix for the prediction points $\mathbf{U}_\star$ are given by

$$\begin{aligned} \mathbb{E}\left[\mathbf{y}_*\right] &= \mathbf{K}_\star^\top \mathbf{K}^{-1}\mathbf{y}\,, \\ \mathrm{Cov}\left[\mathbf{y}_*\right] &= \mathbf{K}_{\star\star} - \mathbf{K}_\star^\top \mathbf{K}^{-1}\mathbf{K}_\star\,. \end{aligned}$$

The hyperparameters $\boldsymbol{\theta}$ are determined by maximum likelihood estimation. From Eq. 2.2 the prior of training outputs is $\mathbf{y} \sim \mathcal{N}(\mathbf{0}, \mathbf{K})$. The likelihood function represents the probability of the observed data $\mathbf{y}$ for a particular model with given parameters. For a single observation $\mathbf{y}$, the likelihood function for the multivariate normal distribution is the same as the PDF evaluated at $\mathbf{y}$:

$$\mathcal{L}(\mathbf{y}|\mathbf{K}) = \frac{1}{(2\pi)^{n/2}\det(\mathbf{K})^{1/2}}\exp\left(-\frac{1}{2}\mathbf{y}^\top \mathbf{K}^{-1}\mathbf{y}\right)\,.$$

The natural logarithm of the likelihood is

$$\ln \mathcal{L}(\mathbf{y}|\mathbf{K}) = -\frac{1}{2}\mathbf{y}^\top \mathbf{K}^{-1}\mathbf{y} - \frac{1}{2}\ln(\det(\mathbf{K})) - \frac{n}{2}\ln(2\pi)\,.$$

The reason to use the log likelihood rather than the likelihood itself is because the log function transforms the product of probabilities into a sum, which is mathematically more convenient and numerically more stable. The natural log likelihood is maximized with respect to the hyperparameters $\boldsymbol{\theta}$. For this purpose, gradient-based optimization has proven to be efficient. Since only $\mathbf{K}$ depends on $\boldsymbol{\theta}$, the gradient of the log marginal likelihood by the use of matrix identities in Appendix A becomes

$$\begin{aligned}\frac{\partial}{\partial \theta_l} \ln \mathcal{L}(\mathbf{y}|\boldsymbol{\theta}) &= \frac{1}{2}\mathbf{y}^\top \mathbf{K}^{-1}\frac{\partial \mathbf{K}}{\partial \theta_l}\mathbf{K}^{-1}\mathbf{y} - \frac{1}{2}\operatorname{tr}\left(\mathbf{K}^{-1}\frac{\partial \mathbf{K}}{\partial \theta_l}\right) \\ &= \frac{1}{2}\operatorname{tr}\left(\left(\boldsymbol{\alpha}\boldsymbol{\alpha}^\top - \mathbf{K}^{-1}\right)\frac{\partial \mathbf{K}}{\partial \theta_l}\right)\end{aligned} \tag{2.3}$$

with $\boldsymbol{\alpha} = \mathbf{K}^{-1}\mathbf{y} = \mathbf{L}_K^{-\top}(\mathbf{L}_K^{-1}\mathbf{y})$ where $\mathbf{L}_K = \text{cholesky}(\mathbf{K})$ denotes the Cholesky decomposition, which factorizes the positive definite matrix $\mathbf{K}$ into a lower triangular matrix $\mathbf{L}_K$ such that $\mathbf{K} = \mathbf{L}_K{\mathbf{L}_K}^\top$.

Using the Einstein summation convention, the derivative of the log marginal likelihood can be expressed as

$$\frac{\partial}{\partial \theta_l} \ln \mathcal{L}(\mathbf{y}|\boldsymbol{\theta}) = \frac{1}{2}\left(\left(\alpha_i\,\alpha_j - \delta_{im}\left[\mathbf{K}^{-1}\right]_{mj}\right)\frac{\partial K_{ji}}{\partial \theta_l}\right)$$

with Kronecker delta $\delta_{im}$.

### 2.2.2 Universal kriging

The theory of universal kriging was introduced by Matheron [87] in the field of geostatistics. Here, the prior is assumed to follow a mean function $m(\mathbf{u})$

$$\hat{y}(\mathbf{u}) \sim \mathcal{GP}(m(\mathbf{u}), k(\mathbf{u}, \mathbf{u}'))\,. \tag{2.4}$$

For this purpose, a set of fixed basis functions $\mathbf{h}(\mathbf{u}) = (h_i(\mathbf{u}), i = 1 \ldots q)$ is defined where the corresponding coefficients $\boldsymbol{\beta} = (\beta_i, i = 1 \ldots q)$ are to be determined. If the prior of the coefficients is assumed to be Gaussian $\boldsymbol{\beta} \sim \mathcal{N}(\mathbf{b}, \boldsymbol{\Sigma}_\beta)$, the prior becomes

$$\hat{y}(\mathbf{u}) \sim \mathcal{GP}(\mathbf{h}(\mathbf{u})^\top\mathbf{b}, k(\mathbf{u}, \mathbf{u}') + \mathbf{h}(\mathbf{u})^\top\boldsymbol{\Sigma}_\beta\mathbf{h}(\mathbf{u}'))\,.$$

The matrices of basis function evaluations at training points $\mathbf{U}$ and prediction points $\mathbf{U}_\star$ are defined as

$$\mathbf{H} = \{H_{ij} = h_i(\mathbf{u}_j), \quad i = 1 \ldots q, \quad j = 1 \ldots n\}\,, \tag{2.5}$$
$$\mathbf{H}_\star = \{H_{\star\,ij} = h_i(\mathbf{u}_{\star\,j}), \quad i = 1 \ldots q, \quad j = 1 \ldots l\}\,. \tag{2.6}$$

The joint distribution of training data $\mathbf{y}$ and prediction data $\mathbf{y}_\star$ is then given by

$$\begin{pmatrix} \mathbf{y} \\ \mathbf{y}_\star \end{pmatrix} \sim \mathcal{N}\left(\begin{pmatrix} \mathbf{H}^\top\mathbf{b} \\ \mathbf{H}_\star^\top\mathbf{b} \end{pmatrix}, \begin{pmatrix} \mathbf{K} + \mathbf{H}^\top\mathbf{\Sigma}_\beta\mathbf{H} & \mathbf{K}_\star + \mathbf{H}^\top\mathbf{\Sigma}_\beta\mathbf{H}_\star \\ \mathbf{K}_\star^\top + \mathbf{H}_\star^\top\mathbf{\Sigma}_\beta\mathbf{H} & \mathbf{K}_{\star\star} + \mathbf{H}_\star^\top\mathbf{\Sigma}_\beta\mathbf{H}_\star \end{pmatrix}\right) . \tag{2.7}$$

By conditioning the prior on the observations, the predictive distribution can be determined as

$$\begin{aligned} \mathbb{E}\left[\mathbf{y}_\star\right] &= \mathbf{H}_\star^\top\mathbf{b} + \left(\mathbf{K}_\star^\top + \mathbf{H}_\star^\top\mathbf{\Sigma}_\beta\mathbf{H}\right)\left(\mathbf{K} + \mathbf{H}^\top\mathbf{\Sigma}_\beta\mathbf{H}\right)^{-1}\left(\mathbf{y} - \mathbf{H}^\top\mathbf{b}\right), \\ \mathrm{Cov}\left[\mathbf{y}_\star\right] &= \mathbf{K}_{\star\star} + \mathbf{H}_\star^\top\mathbf{\Sigma}_\beta\mathbf{H}_\star - \left(\mathbf{K}_\star^\top + \mathbf{H}_\star^\top\mathbf{\Sigma}_\beta\mathbf{H}\right)\left(\mathbf{K} + \mathbf{H}^\top\mathbf{\Sigma}_\beta\mathbf{H}\right)^{-1}\left(\mathbf{K}_\star + \mathbf{H}^\top\mathbf{\Sigma}_\beta\mathbf{H}_\star\right) . \end{aligned}$$

Using the Woodbury identity

$$\left(\mathbf{K} + \mathbf{H}^\top\mathbf{\Sigma}_\beta\mathbf{H}\right)^{-1} = \mathbf{K}^{-1} - \mathbf{K}^{-1}\mathbf{H}^\top\left(\mathbf{\Sigma}_\beta^{-1} + \mathbf{H}\mathbf{K}^{-1}\mathbf{H}^\top\right)^{-1}\mathbf{H}\mathbf{K}^{-1}$$

the expressions become

$$\mathbb{E}\left[\mathbf{y}_\star\right] = \mathbf{K}_\star^\top\mathbf{K}^{-1}\mathbf{y} + \mathbf{R}^\top\bar{\boldsymbol{\beta}} , \tag{2.8}$$

$$\mathrm{Cov}\left[\mathbf{y}_\star\right] = \mathbf{K}_{\star\star} - \mathbf{K}_\star^\top\mathbf{K}^{-1}\mathbf{K}_\star + \mathbf{R}^\top(\mathbf{H}\mathbf{K}^{-1}\mathbf{H}^\top)^{-1}\mathbf{R} \tag{2.9}$$

with $\mathbf{R} = \mathbf{H}_\star - \mathbf{H}\mathbf{K}^{-1}\mathbf{K}_\star$ and $\bar{\boldsymbol{\beta}} = (\mathbf{H}\mathbf{K}^{-1}\mathbf{H}^\top + \mathbf{\Sigma}_\beta{}^{-1})^{-1}\left(\mathbf{H}\mathbf{K}^{-1}\mathbf{y} + \mathbf{\Sigma}_\beta{}^{-1}\mathbf{b}\right)$.
Assuming a non-informative prior on the coefficients ($\mathbf{\Sigma}_\beta{}^{-1} \rightarrow 0$), the vector $\bar{\boldsymbol{\beta}}$ becomes

$$\bar{\boldsymbol{\beta}} = (\mathbf{H}\mathbf{K}^{-1}\mathbf{H}^\top)^{-1}\mathbf{H}\mathbf{K}^{-1}\mathbf{y}$$

and the expressions are independent of the coefficient vector $\boldsymbol{\beta}$ and its prior. Compared to the simple kriging equations, only the basis functions $\mathbf{h}(\mathbf{u})$ have to be specified to apply the universal kriging equations. In the case of a constant scalar basis function $h(\mathbf{u}) = 1$, the equations become

$$\begin{aligned} \mathbb{E}\left[\mathbf{y}_\star\right] &= \mu\,\mathbf{I} + \mathbf{K}_\star^\top\mathbf{K}^{-1}(\mathbf{y} - \mu\,\mathbf{I}) , \\ \mathrm{Cov}\left[\mathbf{y}_\star\right] &= \mathbf{K}_{\star\star} - \mathbf{K}_\star^\top\mathbf{K}^{-1}\mathbf{K}_\star + \mathbf{R}^\top(\mathbf{H}\mathbf{K}^{-1}\mathbf{H}^\top)^{-1}\mathbf{R} \end{aligned}$$

where $\mu = \frac{\mathbf{I}^\top\mathbf{K}^{-1}\mathbf{y}}{\mathbf{I}^\top\mathbf{K}^{-1}\mathbf{I}}$. This case is known as ordinary kriging, i. e. kriging with unknown constant mean.

Analogously to simple kriging, the marginal likelihood is used to estimate the hyperparameters. From Eq. 2.7 the prior of training outputs is $\mathbf{y} \sim \mathcal{N}(\mathbf{H}^\top\mathbf{b}, \mathbf{K}+\mathbf{H}^\top\mathbf{\Sigma}_\beta\mathbf{H})$. The likelihood function is then

$$\mathcal{L}(\mathbf{y}|\mathbf{K}) = \frac{1}{(2\pi)^{\frac{n}{2}}\det(\mathbf{K}+\mathbf{H}^\top\mathbf{\Sigma}_\beta\mathbf{H})^{\frac{1}{2}}}\exp\left(-\frac{1}{2}\left(\mathbf{H}^\top\mathbf{b}-\mathbf{y}\right)^\top\left(\mathbf{K}+\mathbf{H}^\top\mathbf{\Sigma}_\beta\mathbf{H}\right)^{-1}\left(\mathbf{H}^\top\mathbf{b}-\mathbf{y}\right)\right).$$

The natural logarithm of the likelihood is

$$\begin{aligned}\ln\mathcal{L}(\mathbf{y}|\mathbf{K}) = &-\frac{1}{2}\left(\mathbf{H}^\top\mathbf{b}-\mathbf{y}\right)^\top\left(\mathbf{K}+\mathbf{H}^\top\mathbf{\Sigma}_\beta\mathbf{H}\right)^{-1}\left(\mathbf{H}^\top\mathbf{b}-\mathbf{y}\right)\\ &-\frac{1}{2}\ln\left(\det(\mathbf{K}+\mathbf{H}^\top\mathbf{\Sigma}_\beta\mathbf{H})\right)-\frac{n}{2}\ln(2\pi).\end{aligned}$$

Assuming again a non-informative prior on the coefficients ($\mathbf{\Sigma}_\beta^{-1} \to 0$) and assuming $\mathbf{b} = \mathbf{0}$ without loss of generality, the log marginal likelihood can be simplified [113] and becomes

$$\begin{aligned}\ln\mathcal{L}(\mathbf{y}|\mathbf{K}) = &-\frac{1}{2}\mathbf{y}^\top\mathbf{K}^{-1}\mathbf{y}-\frac{1}{2}\ln(\det(\mathbf{K}))-\frac{1}{2}\ln(\det(\mathbf{A}))\\ &+\frac{1}{2}\mathbf{y}^\top\left(\mathbf{K}^{-1}\mathbf{H}^\top\mathbf{A}^{-1}\mathbf{H}\mathbf{K}^{-1}\right)\mathbf{y}-\frac{n-m}{2}\ln(2\pi)\end{aligned}$$

with $\mathbf{A} = \mathbf{H}\mathbf{K}^{-1}\mathbf{H}^\top$.

## 2.3 Principal component regression

As a fundamental method, PCA is widely used for dimensionality reduction and feature extraction. While the previous section (Section 2.2) focused on scalar data, PCA can also be applied to 2D field data $\mathbf{M} = \{\mathbf{M}_i, i = 1\dots n\}$, where each sample $\mathbf{M}_i \in \mathbb{R}^{K\times L}$ represents a structured grid, such as meteorological data with $K$ latitudinal and $L$ longitudinal points. PCA identifies dominant patterns by transforming standardized data into orthogonal principal components, ranked by the variance they capture. This chapter first outlines the fundamentals of PCA and its application to standardized field data. Based on PCA results, PCR is then introduced as a surrogate modeling approach to approximate the training data $\{\mathbf{U}, \mathbf{M}\}$.

### 2.3.1 Principal component analysis

The training data $\mathbf{M} = \{\mathbf{M}_i, i = 1\dots n\}$ are standardized using the mean

$$\mu^{kl} = \frac{1}{n}\sum_{i=1}^{n} M_i^{kl} \tag{2.10}$$

and the standard deviation

$$\sigma^{kl} = \sqrt{\frac{1}{n}\sum_{i=1}^{n}\left(M_i^{kl} - \mu^{kl}\right)^2} \tag{2.11}$$

over all $n$ training points at each grid point $(k, l)$, respectively. The standardized fields $\tilde{\mathbf{M}}_i$ of all $i = 1 \ldots n$ training points can then be expressed as

$$\tilde{M}_i^{kl} = \frac{1}{\sigma^{kl}}\left(M_i^{kl} - \mu^{kl}\right)$$

with zero mean and unit standard deviation at each grid point $(k, l)$.

The matrix components of $\tilde{\mathbf{M}}_i$ are reshaped into vectors $\tilde{\mathbf{m}}_i$ where $\tilde{\mathbf{m}}_i \in \mathbb{R}^{KL}$. The covariance matrix

$$\mathbf{C} = \frac{1}{KL - 1}\sum_{i=1}^{n}\tilde{\mathbf{m}}_i\tilde{\mathbf{m}}_i^{\top}$$

is then computed using all $n$ training points. The eigenvalues and eigenvectors of covariance matrix $\mathbf{C}$ are computed using eigenvalue decomposition for the eigenvalue equation

$$\mathbf{C}\mathbf{p}_m = \lambda_m \mathbf{p}_m$$

where $\lambda_m$ is an eigenvalue and $\mathbf{p}_m$ the corresponding eigenvector. The eigenvectors are sorted in descending order of their corresponding eigenvalues. The top $P$ eigenvectors form the basis of the principal components. These vectors are transformed back into matrix form, resulting in the principal fields $\mathbf{P}_m$ $(m = 1 \ldots P)$, with $P$ principal fields. The fields represent the major variations within the field data.

### 2.3.2 Regression model

The fields $\mathbf{M}$ can now be approximated as a linear combination of the principal fields w. r. t. input parameter vector $\mathbf{u}$

$$\hat{M}^{kl}(\mathbf{u}) \approx \mu^{kl} + \sigma^{kl}\tilde{M}^{kl}(\mathbf{u}) \tag{2.12}$$

$$\text{with} \quad \tilde{M}^{kl}(\mathbf{u}) = \sum_{m=1}^{P} C_m(\mathbf{u}) P_m^{kl} \,,$$

where the coefficients $C_m(\mathbf{u})$ are assumed to be functions w. r. t. the input parameters $\mathbf{u}$ and containing unknown parameters. These parameters can be determined by minimizing the mean square error (MSE) between the surrogate model $\tilde{M}^{kl}(\mathbf{u})$ and the training data $\tilde{M}_i^{kl}$

$$\mathrm{MSE} = \left( \sum_{i=1}^{n} \sum_{k=1}^{K} \sum_{l=1}^{L} \left( \tilde{M}^{kl}(\mathbf{u}_i) - \tilde{M}_i^{kl} \right)^2 \right) . \tag{2.13}$$

In the case of a linear model for the coefficients $C_m(\mathbf{u})$ w. r. t. the input parameters $\mathbf{u}$

$$C_m(\mathbf{u}) = C_m^0 + \sum_{j=1}^{p} C_m^j u_j \, , \tag{2.14}$$

the coefficients $\mathbf{C}_m = \left(C_m^0, \ldots, C_m^p\right)^\top$ can be determined analytically using the least squares solution as

$$\mathbf{C}_m = \left(\mathbf{U}^\top \mathbf{U}\right)^{-1} \mathbf{U}^\top \mathbf{S}_m \, , \tag{2.15}$$

where $\mathbf{S}_m = (S_{m1}, \ldots, S_{mn})^\top$ denotes the vector of principal component scores for each principal component $m$,

$$S_{mi} = \sum_{k,l} \tilde{M}_i^{kl} p_m^{kl} \, .$$

Other models for the coefficients $C_m(\mathbf{u})$ beyond the traditional linear model (Eq. 2.14) are possible; however, in such cases, the analytical solution (Eq. 2.15) would no longer be applicable.

## 2.4 Model Validation

When constructing surrogate models, it is essential to assess their quality using model validation measures. Model accuracy depends on various factors, e. g. the number of training points, the choice of basis functions and nonlinearities in the physical model. Loeppky, Sacks, and Welch [76] recommended to use $n = 10p$ training points for a $p$-dimensional problem in an initial step. Sequential algorithms can then supplement the base design by adding training points in regions where higher model accuracy is required. However, validating the surrogate model remains essential. For this purpose, the generalization error serves as a measure of how accurately the model predicts outcome values for previously unseen data. To quantify the generalization error, the squared error (SE), also known as the $L^2$-error,

$$\mathrm{SE} = \mathbb{E}\left[(y(\mathcal{X}) - \hat{y}(\mathcal{X}))^2\right]$$

is used. This metric represents the expected squared difference between the original physical model $y$ and the surrogate model prediction $\hat{y}$. When using input parameters following a multivariate PDF, the $L^2$-error can be expressed as

$$\mathrm{SE} = \int_{\mathcal{D}_{\mathbf{x}}} (y(\mathbf{x}) - \hat{y}(\mathbf{x}))^2 \,\mathrm{PDF}_{\mathcal{X}}(\mathbf{x}) \mathrm{d}\mathbf{x}\,.$$

For i. i. d. uniform inputs with $\mathrm{PDF}_{\mathcal{U}}(\mathbf{u}) = 1$ and the support $\mathcal{D}_{\mathcal{U}} = [0,1]^p$, the $L^2$-error is

$$\mathrm{SE} = \int_{\mathcal{D}_{\mathcal{U}}} (y(\mathbf{u}) - \hat{y}(\mathbf{u}))^2 \,\mathrm{d}\mathbf{u}\,.$$

The SE value is generally not known analytically, since $y$ is only known for a finite set of evaluations, particularly in the case of complex computer models. Therefore, the generalization error can be estimated with a validation data set with $n_{\mathrm{val}}$ validation points $\{(\mathbf{u}_{\mathrm{val},i}, y_{\mathrm{val},i}),\ i = 1 \ldots n_{\mathrm{val}}\}$ obtained from evaluations of the computer model. The validation points are obtained by MC sampling with respect to the parameter PDFs. The root mean square error (RMSE) and the normalized mean square error (NMSE) become

$$\mathrm{RMSE} = \sqrt{\frac{1}{n_{\mathrm{val}}} \sum_{i=1}^{n_{\mathrm{val}}} (y_{\mathrm{val},i} - \hat{y}(\mathbf{u}_{\mathrm{val},i}))^2}\,,$$

$$\mathrm{NMSE} = \frac{1}{\sigma^2_{y_{\mathrm{val}}}} \frac{1}{n_{\mathrm{val}}} \sum_{i=1}^{n_{\mathrm{val}}} (y_{\mathrm{val},i} - \hat{y}(\mathbf{u}_{\mathrm{val},i}))^2\,.$$

Here,

$$\sigma^2_{y_{\mathrm{val}}} = \frac{1}{n_{\mathrm{val}}} \sum_{i=1}^{n_{\mathrm{val}}} \left(y_{\mathrm{val},i} - \overline{y}\right)^2\,,$$

$$\overline{y} = \frac{1}{n_{\mathrm{val}}} \sum_{i=1}^{n_{\mathrm{val}}} y_{\mathrm{val},i}$$

are the variance and the mean of $\mathbf{y}_{\mathrm{val}}$, respectively. Conversely, goodness-of-fit measures like the coefficient of determination $R^2$, where the same points are used for training and validation, do not account for overfitting and should therefore be avoided.

While RMSE provides insights into absolute error values, NMSE offers a dimensionless measure facilitating better comparison between different output variables. Model accuracy is deemed high if NMSE values approach zero and low if they approach one. Values are inherently non-negative and should not exceed one, as this would indicate that the covariance between the surrogate model and the data surpasses the data's variance. The NMSE error is used since normalization allows comparison between different physical quantities, i. e. different scales. The interpretation of the NMSE can become problematic when the QoI values exhibit little variation, resulting in a very small variance $\sigma^2_{y_{\mathrm{val}}}$. In such cases, the RMSE should be considered.

Because of high computation cost, using a separate validation set that is not used for model training is not effective. Therefore, cross-validation techniques, such as leave-one-out validation or leave-$k$-out validation, can be applied, where seperate surrogate models are computed using subsets of the training data and validated with data which was not used for training. The validation errors for leave-$k$-out-validation can be formulated as

$$\mathrm{RMSE} = \sqrt{\frac{1}{n}\sum_{i=1}^{n}(y_i - \hat{y}_{\setminus K(i)}(\mathbf{u}_i))^2}\,, \tag{2.16}$$

$$\mathrm{NMSE} = \frac{1}{\sigma_y^2}\frac{1}{n}\sum_{i=1}^{n}(y_i - \hat{y}_{\setminus K(i)}(\mathbf{u}_i))^2\,, \tag{2.17}$$

where $\hat{y}_{\setminus K(i)}$ denotes the surrogate model trained on all $n$ training points except those in the subset $K(i)$, which contains the $i$-th point. When using $k$ points per subset, the data is partitioned into disjoint subsets $K = ((1,\ldots,k),(k+1,\ldots,2k),\ldots,(n-k+1,\ldots,n))$, with each point assigned to a subset accordingly: for example, $K(1) = \ldots = K(k) = (1,\ldots,k)$, $K(k+1) = \ldots = K(2k) = (k+1,\ldots,2k)$, and so on. In the case of leave-2-out cross-validation, the subsets take the form $K = ((1,2),(3,4)\ldots(n-1,n))$. The variance of evaluations $y_i$ is denoted by $\sigma_y^2$. Since, in principle, a separate surrogate model must be constructed for each subset of training points in cross-validation, certain surrogate modeling techniques provide analytical expressions that enable computationally efficient cross-validation without the need to recompute the entire surrogate model for every subset (see e. g. Dubrule [18] for GPR).

For field data, the validation measures are adapted accordingly. The accuracy of surrogate models depends on the underlying PCR analysis, including factors such as the number of principal fields and the choice of the coefficient ansatz function. The corresponding error measures are defined as

$$\text{RMSE} = \sqrt{\frac{1}{KLn}\sum_{i=1}^{n}\sum_{k=1}^{K}\sum_{l=1}^{L}(M_i^{kl} - \hat{M}_{\backslash K(i)}^{kl}(\mathbf{u}_i))^2}\,, \tag{2.18}$$

$$\text{NMSE} = \frac{1}{\sigma_M^2}\frac{1}{KLn}\sum_{i=1}^{n}\sum_{k=1}^{K}\sum_{l=1}^{L}(M_i^{kl} - \hat{M}_{\backslash K(i)}^{kl}(\mathbf{u}_i))^2. \tag{2.19}$$

Here, $\sigma_M^2$ denotes the variance of the field data $\mu^{kl}$ across all grid points $(k, l)$ and $\mathbf{M}_{\backslash K(i)}(\mathbf{u})$ represents the surrogate model derived from all $n$ training points except those within set $K(i)$ containing the $i$-th point.

## 2.5 Global sensitivity analysis

To quantify the relative influence of uncertain model parameters on QoIs, GSA can be employed. In this work, variance-based sensitivity analysis [120], often referred to as the Sobol' method [125], is used as a form of GSA to decompose the variance of a scalar model output into fractions attributable to individual inputs or sets of inputs. From the PIT (Section 2.1.3), the input space can be described by the input random variables $(U_1, U_2 \ldots U_p)$, which are i. i. d. and uniformly distributed.

Variance-based sensitivity analysis focuses on decomposing the variance of the output random variable $Y = y(U_1, U_2, \ldots, U_p)$

$$V = \text{Var}[Y]$$

into contributions from each input variable or combinations of variables. To define the sensitivity indices, the unique functional ANOVA decomposition of any integrable function on $[0, 1]^p$ into a sum of elementary functions is used [125]

$$Y = y_0 + \sum_{i=1}^{p} y_i(U_i) + \sum_{i<j}^{p} y_{ij}(U_i, U_j) + \cdots + y_{12\ldots p}(U_1, U_2, \ldots, U_p)\,,$$

where $y_0$ is a constant and $y_i, y_{ij}$ etc. are functions of the decomposition w.r.t. the random inputs, respecting the orthogonality condition

$$\mathbb{E}[y_J(U_J)] = 0$$

for all subsets $J$ of input parameters (e. g. for $J = \{1,2\} : y_J \triangleq J_{12}$, $U_J \triangleq (U_1, U_2)$).

This leads to definitions of the terms of the functional decomposition in terms of conditional expected values,

$$\begin{aligned} y_0 &= \mathbb{E}[Y]\,, \\ y_i(U_i) &= \mathbb{E}[Y \mid U_i] - y_0\,, \\ y_{ij}(U_i, U_j) &= \mathbb{E}[Y \mid U_i, U_j] - y_0 - y_i - y_j \quad \text{etc.} \end{aligned}$$

Here, it becomes evident that $y_i$ is the effect of varying $U_i$ alone (main effect), and $y_{ij}$ in addition includes the effect of varying $U_i$ and $U_j$ simultaneously. This is known as a second-order interaction. Higher-order terms have analogous definitions. Further assuming that $y(U_1, U_2 \ldots U_p)$ is square-integrable, the functional decomposition may be squared and integrated

$$\int y^2 \,\mathrm{d}U_1 \cdots \mathrm{d}U_p = y_0^2 + \sum_{s=1}^{p} \sum_{1 \leq i_1 < i_2 < \cdots < i_s \leq p} \int y^2_{i_1 i_2 \ldots i_s} \,\mathrm{d}U_{i_1} \cdots \mathrm{d}U_{i_s}$$

and can be expressed as

$$\mathrm{Var}[Y] = \sum_{i=1}^{p} V_i + \sum_{i<j}^{d} V_{ij} + \cdots + V_{12\ldots p}\,,$$

where

$$\begin{aligned} V_i &= \mathrm{Var}_{U_i}\left[\mathbb{E}_{U_{\setminus i}}[Y \mid U_i]\right]\,, \\ V_{ij} &= \mathrm{Var}_{U_{ij}}\left[\mathbb{E}_{U_{\setminus ij}}[Y \mid U_i, U_j]\right] - V_i - V_j \quad \text{etc.} \end{aligned}$$

The notation $U_{\setminus i}$ indicates the set of all variables except $U_i$, whereas $U_{\setminus ij}$ indicates the set of all variables except $U_i$ and $U_j$ etc. The above variance decomposition shows how the variance of the model output can be decomposed into terms attributable to each input, as well as the interaction effects between them. Together, all terms sum to the total variance of the model output.

The main effect index (first-order sensitivity index) $S_i$ measures the direct contribution of input $U_i$ and is defined as

$$S_i = \frac{V_i}{V}\,,$$

where $V_i$ represents the contribution to the output variance due to $U_i$ alone. The total effect index $S_{Ti}$ measures the total contribution of input $U_i$, including both its direct effect and all interaction effects with other inputs, and is defined as

$$S_{Ti} = \sum_{J \supseteq \{i\}} S_J = \frac{\sum_{J \supseteq \{i\}} V_J}{\mathrm{Var}[Y]} = 1 - \frac{V_{\setminus i}}{V},$$

where $V_{\setminus i} = \mathrm{Var}_{U_{\setminus i}}\left[\mathbb{E}_{U_i}[Y|U_{\setminus i}]\right]$ is the variance of the output $Y$ excluding the effect of $U_i$.

Estimation of the sensitivity indices can be done using MC simulations (Section 2.1.4) or alternative methods. One prominent approach is the Fourier Amplitude Sensitivity Test (FAST) [120]. This approach estimates the variance-based Sobol indices by transforming the multidimensional input space into a single-dimensional space using a sinusoidal trajectory. This transformation allows for efficient exploration of the input space and reduces the computational cost. For this purpose, each input variable $U_i$ is assigned a unique frequency $\omega_i$, and the model output $Y$ is evaluated along a curve defined by these frequencies. The relationship between the input variables and the model output can then be analyzed using Fourier decomposition. The first-order Sobol index $S_i$ is computed as

$$S_i = \frac{A_{\omega_i}^2}{V},$$

where $A_{\omega_i}$ is the amplitude of the Fourier component corresponding to the frequency $\omega_i$. FAST can also be used and modified to compute higher-order Sobol indices and total Sobol indices $S_{Ti}$. The accuracy of computed indices depends on appropriately selecting the frequencies $\omega_i$ to avoid aliasing and ensure sufficient spectral resolution.

# 3 Extended concepts in surrogate modeling and optimization

This section presents a framework to advance fundamental methodologies, supporting the primary objective of quantifying uncertainties and optimizing model parameters in simulations of the WAM. The framework integrates non-uniform input parameter distributions and adapts sampling methods based on the underlying PDFs. For universal kriging, the gradient of the log marginal likelihood is computed and utilized. Additionally, two multi-objective optimization concepts are introduced as a systematic framework.

## 3.1 Surrogate models for transformed input parameter spaces

Surrogate modeling techniques in UQ often rely on predefined parameter bounds, where sampling methods (Section 2.1.4), particularly space-filling designs, are applied within the original parameter space, ignoring the probability distributions which these parameters are assumed to follow, and overlooking the need for denser sampling in certain regions of interest. To address this, this section introduces a sophisticated framework for adapting surrogate modeling methods to scenarios where input parameters follow non-uniform distributions [158].

### 3.1.1 Input space transformation

The objective of this work is to construct a surrogate model over the input space with sufficient accuracy, placing particular emphasis on achieving higher precision in critical regions. For instance, in reliability engineering, enhanced accuracy is often crucial in areas near the failure domain [73]. In this study, the focus is on model parameters described by PDFs, which define regions of the parameter space with varying probabilities. The model validation strategy involves sampling validation points according to the joint PDF of these parameters, ensuring that the surrogate model attains the highest accuracy in high-probability regions. This approach, emphasized in the literature [2, 133, 154], is particularly beneficial for analyses such as GSA or other applications that rely on sampling based on input distributions.

However, concentrating training points in regions of higher probability density results in an experimental design with non-uniform space-filling properties, which can present challenges for certain surrogate modeling methods. To address this, it is beneficial to consider an i. i. d. uniform input space. A system with $p$ input parameters and an associated joint PDF is considered. By applying PIT (Section 2.1.3), specifically the Rosenblatt transformation, the physical input space – with its dependence structure captured by a joint PDF – is mapped to an i. i. d. uniform input space. If the inputs are independent, the joint PDF can be factorized into $p$ individual PDFs, allowing the Rosenblatt transformation to be replaced by independent, parameter-wise transformations using the respective PPFs. In this resulting $p$-dimensional unit hypercube, all regions have equal probability, enabling space-filling sampling techniques such as LHS to effectively align the distribution of training points with the probability distribution of the input space.

This uniform input space offers additional advantages, as validation techniques can be directly applied using MC sampling within these uniform boundaries, and variance-based sensitivity analysis can be conducted without requiring further transformations or adjustments to account for non-uniform probability distributions. Once training points are generated in the uniform input space, they are mapped back to the physical input space to obtain the physical parameters used in the computational model simulations.

Various surrogate modeling methods allow for the specification of trend functions as basis functions, ranging from multivariate linear trends to higher-order polynomials, depending on the problem's dimensionality and the number of available training points. The complexity of these trend functions must strike a balance between flexibility and the risk of overfitting. However, when the experimental design is established in the uniform input space, where the surrogate models are constructed, defining trend functions with respect to these uniform parameters is not optimal. This is because the physical relationships that trend functions aim to capture are generally more directly associated with the physical input parameters. Therefore, the input space transformation should be accounted for when defining trend functions, as will be demonstrated for GPR and PCR.

### 3.1.2 Implications for Gaussian process regression

Universal kriging provides a flexible and powerful approach for incorporating explicit basis functions with respect to the input parameters as underlying trends of the system. In conventional studies, the kriging equations (Eq. 2.8, Eq. 2.9) are typically expressed directly in terms of the physical input parameters. However, in this study, the surrogate modeling procedure is conducted in the i. i. d. uniform input space (see Section 3.1.1).

Since the trend functions $\mathbf{f}(\mathbf{x})$ are initially defined with respect to the physical input parameters $\mathbf{x}$, they must be transformed into the i. i. d. uniform input space using the inverse Rosenblatt transformation $\mathcal{T}_{\text{ros}}^{-1}$. The transformed basis functions are given by

$$h(\mathbf{u}) = f(\mathcal{T}_{\text{ros}}^{-1}(\mathbf{u})) \tag{3.1}$$

with respect to the uniform parameters $\mathbf{u}$. These basis functions $\mathbf{h}(\mathbf{u})$ can then be used in the universal kriging equations (Eq. 2.4), specifically for computing $\mathbf{H}$ and $\mathbf{H}_\star$ (Eq. 2.5 and Eq. 2.6).

In the one-dimensional case, the transformation simplifies to the PPF, such that $x = \text{PPF}(u)$ for the given input parameter. Consequently, the transformed basis function becomes $h(u) = f(\text{PPF}(u))$. For the specific case of the linear basis function $f(x) = x$, this transformation results in $h(u) = \text{PPF}(u)$. If the input parameters are uncorrelated, i. e. $\{\rho_{X_i,X_j} = 0 \quad \forall i, j \in \{1, \ldots, p\},\ i \neq j\}$, the transformation reduces to independent functions $\{h(u_i) = f(\text{PPF}(u_i)),\ i = 1 \ldots p\}$ for all input parameters.

All equations can also be formulated with respect to the physical parameters $\mathbf{x}$ instead of the uniform parameters $\mathbf{u}$. In this case, the original, non-transformed definition of trend functions is retained. However, this results in a non-stationary kernel function, as the kernel functions are assumed to be stationary with respect to the i. i. d. uniform variables. Consequently, the formulation of the kernel (Eq. 2.1) takes the form

$$k(\mathbf{x}, \mathbf{x}') = \theta_0 \exp\left( - \sum_{j=1}^{p} \left( \frac{|\mathcal{T}_{\text{ros}}(\mathbf{x})_j - \mathcal{T}_{\text{ros}}(\mathbf{x}')_j|}{\theta_j} \right)^2 \right).$$

It is important to emphasize that defining stationary kernel functions in the i. i. d. uniform parameter space is preferable to assuming stationarity in the physical parameter space. In the physical parameter space, the density of training points corresponds to the joint PDF of the input parameters. Consequently, training points are sparse in the tails of the PDF, where a larger kernel length scale is desirable, while in high-density regions (i. e. closer to the mean), a smaller length scale is more appropriate. This adaptation

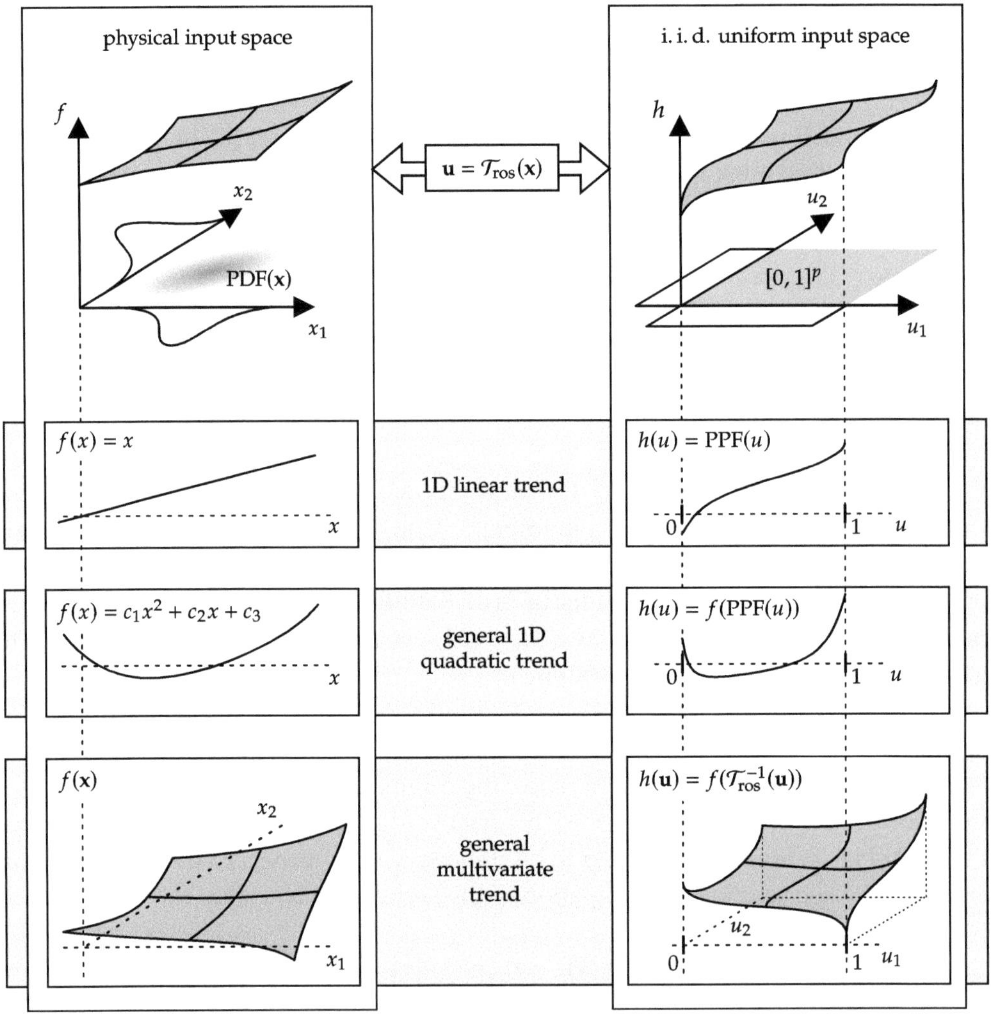


**Figure 3.1:** Visualization of model functions (top) and selected trend functions (three rows below) in the physical (left) and i. i. d. uniform (right) parameter space. The Rosenblatt transformation $\mathcal{T}_{\text{ros}}$ (PPF in the case of independent variables) defines the relationship between both parameter spaces.

is naturally achieved by defining a stationary kernel function in the i. i. d. uniform input space, where the density of training points is uniform. Figure 3.1 illustrates the transformation of basis functions. Related ideas involving transformations to ensure stationarity in the transformed input space have been explored in previous studies [121, 123]. Similarly, other studies have investigated the use of non-stationary kernel functions [37, 147], typically based on variations in function smoothness. However, these methods did not account for the joint PDFs of the input parameters and the corresponding density of training points in the experimental design, as considered in this work.

Hyperparameter optimization can be computationally demanding for high-dimensional problems with a large number of training points. For simple kriging, the gradient of the log marginal likelihood has been previously shown in Eq. 2.3. For universal kriging [113], the log marginal likelihood is given by

$$\log p(\mathbf{y}|\mathbf{U},\boldsymbol{\theta}) = -\frac{1}{2}\mathbf{y}^\top\mathbf{K}^{-1}\mathbf{y} + \frac{1}{2}\mathbf{y}^\top\mathbf{C}\mathbf{y} - \frac{1}{2}\log|\mathbf{K}| - \frac{1}{2}\log|\mathbf{A}| - \frac{n-m}{2}\log 2\pi\,,$$

where $\mathbf{A} = \mathbf{H}\mathbf{K}^{-1}\mathbf{H}^\top$, $\mathbf{C} = \mathbf{K}^{-1}\mathbf{H}^\top\mathbf{A}^{-1}\mathbf{H}\mathbf{K}^{-1}$ and $m = \mathrm{rank}(\mathbf{H}^\top)$ (see Section 2.2 for other quantities). The derivation of its gradient, as presented in Appendix B, yields

$$\frac{\partial}{\partial\theta_l}\log p(\mathbf{y}|\mathbf{U},\boldsymbol{\theta}) = \frac{1}{2}\mathrm{tr}\left(\left(\rho - \xi - \xi^\top + \xi\varepsilon^\top + (\varepsilon - \mathbb{1})\mathbf{K}^{-1}\right)\frac{\partial\mathbf{K}}{\partial\theta_l}\right)\,, \tag{3.2}$$

where

$$\begin{aligned}
\rho &= \alpha\alpha^\top\,, & \alpha &= \mathbf{K}^{-1}\mathbf{y} = \mathbf{L}_K^{-\top}(\mathbf{L}_K^{-1}\mathbf{y})\,,\\
\varepsilon &= \gamma\eta\,, & \gamma &= \mathbf{K}^{-1}\mathbf{H}^\top = \mathbf{L}_K^{-\top}(\mathbf{L}_K^{-1}\mathbf{H}^\top)\,,\\
\xi &= \varepsilon\rho\,, & \eta &= (\mathbf{H}\mathbf{K}^{-1}\mathbf{H}^\top)^{-1}\mathbf{H} = \mathbf{L}_\eta^{-\top}(\mathbf{L}_\eta^{-1}\mathbf{H})\,,
\end{aligned}$$

with $\mathbf{L}_K = \mathrm{cholesky}(\mathbf{K})$ and $\mathbf{L}_\eta = \mathrm{cholesky}(\mathbf{H}\mathbf{K}^{-1}\mathbf{H}^\top)$. The Cholesky decomposition is applied to efficiently compute the inverse. The derivative in Eq. 3.2, along with the given quantities, is utilized when the proposed transformed trend functions (Eq. 3.1) are incorporated into universal kriging, and gradient-based hyperparameter estimation is performed.

### 3.1.3 Implications for principal component regression

Traditionally, the PCR model is linear in terms of the coefficients of the principal components. As a result, it benefits from well-established properties of linear regression, including an analytic solution for the coefficients (see Eq. 2.15). However, in this study, a transformed input parameter space is considered, where both sampling and surrogate modeling methods are applied. To improve accuracy, the input space transformation is incorporated into the PCR model, albeit at the cost of losing these advantageous properties.

The coefficients of the principal fields are generally expressed as functions of the physical model parameters. To account for the input space transformation, the ansatz for the coefficients $C_m(\mathbf{u})$ (Eq. 2.14) is adapted accordingly. Specifically, assuming a linear ansatz in the physical parameters $\mathbf{x}$, the coefficient equations take the form

$$C_m(\mathbf{u}) = C_m^0 + \sum_{j=1}^{p} C_m^j \mathcal{T}_{\mathrm{ros}}^{-1}(\mathbf{u})_j \, . \tag{3.3}$$

For independent input parameters $\mathbf{x}_j$ ($j = 1 \ldots p$), this expression simplifies to

$$C_m(\mathbf{u}) = C_m^0 + \sum_{j=1}^{p} C_m^j \, \mathrm{CDF}^{-1}(u_p) \, ,$$

where the coefficients are expressed with respect to the i. i. d. uniform input parameters $\mathbf{u} = (u_1, \ldots, u_p)^\top$. The optimal parameters $C_m^j$ are determined by minimizing the MSE (Eq. 2.13).

### 3.1.4 Test cases

The proposed method of constructing surrogate models for transformed input parameter spaces in Section 3.1 is validated with several benchmark problems. For this purpose, GPR (Section 3.1.2) is considered, where the derived equations are employed for hyperparameter estimation. It should be noted that a separate validation for PCR would be sensible, however, for GPR, the validation offers numerous advantages, such as available established benchmark problems for scalar-valued problems in UQ, the lower computation time and easier interpretability. Furthermore, the derived equations for the gradient of the log marginal likelihood (Appendix B) can be directly employed and investigated as well.

#### Benchmark problems

The benchmark problems considered in this study, which are widely used in UQ, are listed in Table 3.1. PDFs are assigned to the input parameters. Each investigated

**Table 3.1:** Benchmark functions with their mathematical expressions, number of input dimensions, PDFs and parameter correlations.

| # | equation | dim. | input parameter PDFs[1] $f_{X_i}$ |
|---|---|---|---|
| 1 | Oakley & O'Hagan [101]<br>$f(x) = 5 + x + \cos(x)$ | 1 | $X \sim \mathcal{N}(0,4)$ |
| 2 | Lognormal Ratio [20]<br>$f(\mathbf{x}) = \frac{x_1}{x_2}$ | 2 | $X_{1,2} \sim \mathcal{LN}(1,0.5)$<br>$\rho_{X_1,X_2} = 0.3$ |
| 3 | Webster et al. [143]<br>$f(\mathbf{x}) = x_1^2 + x_2^3$ | 2 | $X_1 \sim \mathcal{U}(1,10)$<br>$X_2 \sim \mathcal{N}(2,1)$ |
| 4 | Short Column [20]<br>$f(\mathbf{x}) = 1 - \frac{4}{1125}\frac{x_2}{x_1} - \frac{1}{5625}\left(\frac{x_3}{x_1}\right)^2$ | 3 | $X_1 \sim \mathcal{LN}(5,0.5)$<br>$X_2 \sim \mathcal{N}(2000,400)$<br>$X_3 \sim \mathcal{N}(500,100)$<br>$\rho_{X_2,X_3} = 0.5$ |
| 5 | Cantilever Beam [20]<br>$f(\mathbf{x}) = \frac{5 \cdot 10^5}{x_1}\sqrt{\left(\frac{x_2}{16}\right)^2 + \left(\frac{x_3}{4}\right)^2}$ | 3 | $X_1 \sim \mathcal{N}(2.9e7, 1.45e6)$<br>$X_2 \sim \mathcal{N}(1000,100)$<br>$X_3 \sim \mathcal{N}(500,100)$ |
| 6 | Borehole [47, 94]<br>$f(\mathbf{x}) = \frac{2\pi x_3(x_4 - x_6)}{\ln\left(\frac{x_2}{x_1}\right)\left(1 + \frac{2x_7x_3}{\ln(x_2/x_1)x_1^2x_8} + \frac{x_3}{x_5}\right)}$ | 8 | $X_1 \sim \mathcal{N}(0.1, 0.0162)$<br>$X_2 \sim \mathcal{LN}(3700,4890)$<br>$X_3 \sim \mathcal{U}(63\,070, 115\,600)$<br>$X_4 \sim \mathcal{U}(990,1110)$<br>$X_5 \sim \mathcal{U}(63.1,116)$<br>$X_6 \sim \mathcal{U}(700,820)$<br>$X_7 \sim \mathcal{U}(1120,1680)$<br>$X_8 \sim \mathcal{U}(9\,855, 12\,045)$ |
| 7 | Steel Column [21]<br>$f(\mathbf{x}) = x_1 - \frac{P}{2x_5x_6} - \frac{x_8PE_b}{x_5x_6x_7(E_b - P)},$<br>$P = x_2 + x_3 + x_4,$<br>$E_b = \frac{8\pi^2}{9 \cdot 10^8}x_5x_6x_7^2x_9$ | 9 | $X_1 \sim \mathcal{LN}(400,35)$<br>$X_2 \sim \mathcal{N}(5e5,5e4)$<br>$X_{3,4} \sim \mathcal{G}(6e5,9e4)$<br>$X_5 \sim \mathcal{LN}(300,3),$<br>$X_6 \sim \mathcal{LN}(20,2)$<br>$X_7 \sim \mathcal{LN}(300,5)$<br>$X_8 \sim \mathcal{N}(30,10)$<br>$X_9 \sim \mathcal{W}(2.1e5,4200)$ |

| | | | |
|---|---|---|---|
| 8 | Sulfur Model [130]<br>$f(\mathbf{x}) = -5.488 \cdot 10^{-9}$<br>$\cdot\, x_1^2 x_2 x_3^2 x_4 x_5 x_6 x_7 x_8 x_9$ | 9 | $X_1 \sim \mathcal{LN}(0.76, 0.152)$<br>$X_2 \sim \mathcal{LN}(0.39, 0.039)$<br>$X_3 \sim \mathcal{LN}(0.85, 0.085)$<br>$X_4 \sim \mathcal{LN}(0.3, 0.09)$<br>$X_5 \sim \mathcal{LN}(5.0, 2.0)$,<br>$X_6 \sim \mathcal{LN}(1.7, 0.34)$<br>$X_7 \sim \mathcal{LN}(71.0, 10.65)$<br>$X_8 \sim \mathcal{LN}(0.5, 0.25)$<br>$X_9 \sim \mathcal{LN}(5.5, 2.75)$ |
| 9 | Oakley & O'Hagan [103]<br>$f(\mathbf{x}) = \mathbf{a}_1^T \mathbf{x} + \mathbf{a}_2^T \sin(\mathbf{x}) + \mathbf{a}_3^T \cos(\mathbf{x})$<br>$+ \mathbf{x}^T \mathbf{M} \mathbf{x} \quad (\mathbf{a}_i, \mathbf{M}\ [103])$ | 15 | $X_i \sim \mathcal{N}(0, 1), \quad i = 1 \ldots 15$ |

[1] PDF parameters correspond to mean $\mu$ and standard deviation $\sigma$ for normal $\mathcal{N}$, log-normal $\mathcal{LN}$, Weibull $\mathcal{W}$ and Gumbel $\mathcal{G}$ distributions and to lower and upper limit for uniform $\mathcal{U}$ distributions. $\rho_{X_i,X_j}$ indicates pairwise Pearson correlation coefficient (0 if not stated).

function includes a non-uniform distribution for at least one input dimension, as the PIT would otherwise have no effect on the trend function, and the proposed method would be indistinguishable from conventional approaches.

For each benchmark problem with $p$ input dimensions, a training set of $n = 10p$ points $\mathbf{u}_i$ is generated using maximin LHS (Section 2.1.4). Model function evaluations $y_i = f(\mathbf{x}_i)$ are performed, where the physical input parameters are obtained via the inverse Rosenblatt transformation, $\mathbf{x}_i = \mathcal{T}_{\text{ros}}^{-1}(\mathbf{u}_i)$. Surrogate models are constructed based on the methods described in Section 3.1.2.

The following GPR methods are compared: simple kriging, ordinary kriging, universal kriging with a linear trend, universal kriging with a quadratic trend, universal kriging with a transformed linear trend, and universal kriging with a transformed quadratic trend. A linear trend includes linear terms with respect to all input parameters, while a quadratic trend includes polynomial terms up to the second order. A transformed trend applies the inverse Rosenblatt transformation to the corresponding trend function, as detailed in Section 3.1.1. Transformations are only applied to linear and quadratic trends, as transforming a zero-order trend (simple kriging) or a constant trend (ordinary kriging) would not produce any change. To ensure statistical validity, each surrogate modeling method is applied 10 times for every combination of a surrogate method and a benchmark problem. Additionally, for each experiment, hyperparameter optimization is repeated 20 times with randomized initial hyperparameters, as multiple local optima may exist and hyperparameter optimization is often sensitive to initial values. The total computation time for one experiment includes all 20 hyperparameter optimization runs. Surrogate models are evaluated based on the validation measures described in Section 2.4, using $n_{\text{val}} = 1000$ validation points. For all cases, the mean and standard deviation of the validation errors are computed.

Furthermore, all experiments are conducted both with and without the gradient of the log marginal likelihood, following Eq. 2.3 and Eq. 3.2, to compare computation times. The L-BFGS-B method is used for optimization because it can directly utilize gradient information, the marginal likelihood of a GPR model is typically smooth and differentiable, and bounded constraints are imposed on the hyperparameters, making this method well-suited. When gradient information is not included, it is estimated using a two-point finite difference approximation.

**Results**

Figure 3.2 illustrates the validation errors for all combinations of benchmark functions and surrogate methods, based on 10 experiments, including their mean values and standard deviations. On average, prediction accuracy improves from simple kriging to ordinary kriging to universal kriging, as increased model flexibility allows better adaptation to the underlying function. In universal kriging, the use of linear or quadratic basis functions, or both, enhances prediction accuracy, with quadratic trends generally yielding lower errors. Quadratic basis functions offer greater flexibility, enabling the surrogate model to better capture the structure of the data. However, polynomials of higher degree, particularly in high-dimensional input spaces, result in an exponentially increasing number of basis functions (curse of dimensionality), which can result in overfitting and high generalization errors. The high validation errors observed for benchmark problem #9 in universal kriging with a quadratic trend indicate the effects of overfitting. In a general case, employing quadratic multivariate basis functions for a $p$-dimensional problem results in $p(p+1)/2$ quadratic, $p$ linear, and one constant basis function. For $p = 15$ with $n = 10p$, this corresponds to 136 basis functions for a problem with only 150 training points. In the context of polynomial chaos expansion, Sudret [127] argued that for polynomial fitting the number of training points should be at least 2-3 times as high as the number of basis functions in order to avoid overfitting. If this is not the case, a sparse set of basis functions could be determined, e. g. by means of least-angle regression [60].

The incorporation of transformed basis functions further reduces validation errors compared to the non-transformed case for both linear and quadratic trends in most cases, underscoring the importance of defining trend functions in the physical parameter space rather than in the uniform space. While the transformation significantly improves accuracy compared to the corresponding non-transformed trend, the choice between a transformed linear or transformed quadratic trend remains highly problem-specific, similar to the selection between linear and quadratic trends without transformation.

The magnitude of validation errors varies significantly across the considered benchmark functions due to differences in mathematical formulations, degrees of nonlinearity, dimensionality, and the number of training points. However, these differences are not the focus of this study.

Figure 3.3 illustrates the impact of incorporating transformed basis functions on the surrogate model for the short column function (benchmark problem #4). For visualization, only cross-sections of the hypersurface are shown, where a single input parameter is varied at a time, and the corresponding output is displayed. The function values to be predicted, along with the surrogate model outputs, are presented in both the physical input space and the i. i. d. uniform input space. Since the surrogate models are constructed in the uniform input space, selecting linear or quadratic trend functions without transformation forces the regression model to conform to such a trend in the uniform space space (Fig. 3.3, third and fourth row). This leads to an artifact where, in the physical input space, the surrogate model tends to flatten in the distribution tails and steepen between. This effect does not occur when transformed basis functions are used, as they correctly account for the input space transformation (Fig. 3.3, fifth and sixth row). This example highlights that simple trends are typically more naturally represented in the physical input space.

Figure 3.4 presents the computation times for all benchmark function and surrogate method combinations, including all 10 experiments, with mean values and standard deviations. Computation times are compared with and without the incorporation of the gradient of the log marginal likelihood in hyperparameter estimation. In all cases, incorporating gradient information leads to a significant reduction in computation time, demonstrating its effectiveness in accelerating optimization. The equations in Section 3.1.2 have thus been validated and shown to be applicable across all cases. Since incorporating gradient information only affects the time required to reach a specified optimality criterion at which the optimization is terminated, but does not alter the final optimization results, the NMSE errors (Fig. 3.2) remain identical between the cases with and without gradient information.

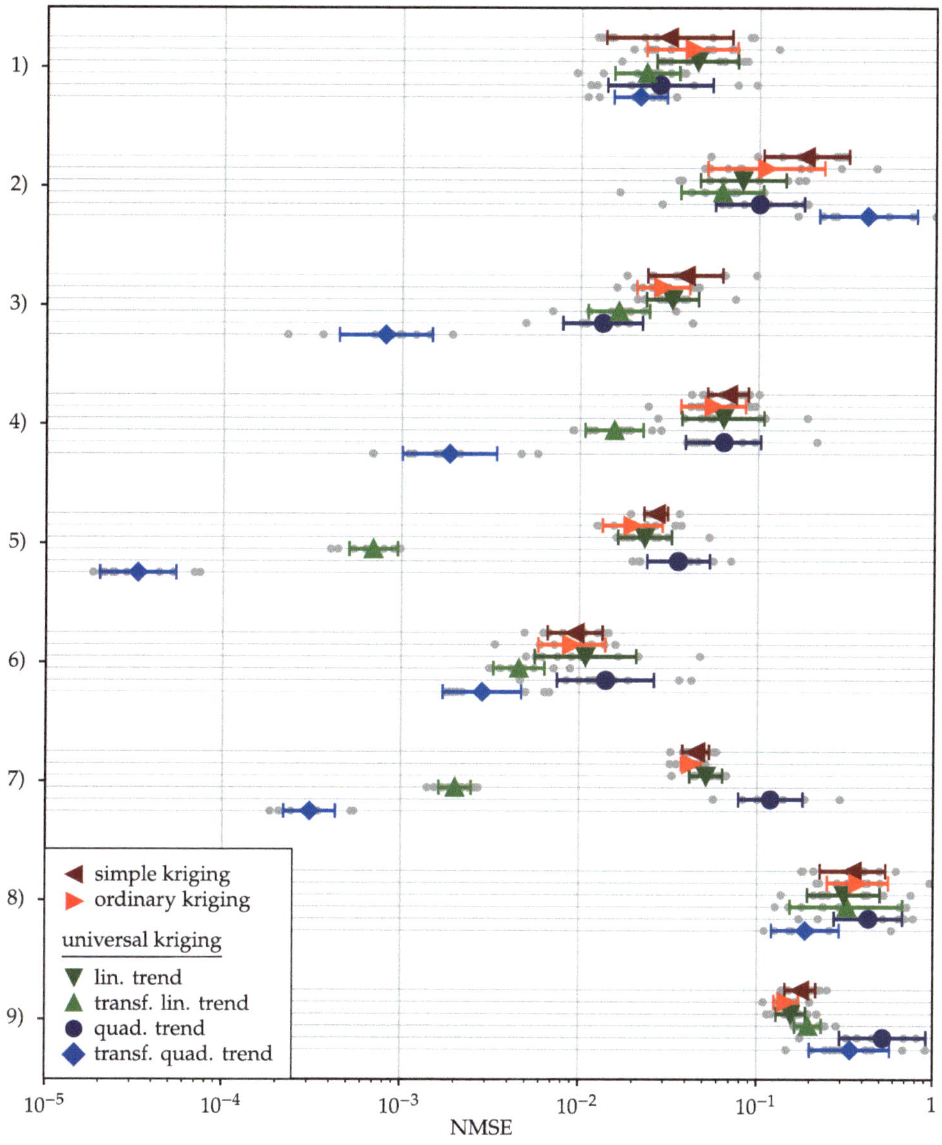


**Figure 3.2:** NMSE for all benchmark problems and investigated GPR methods, including mean values and standard deviations across 10 experiments for each case.

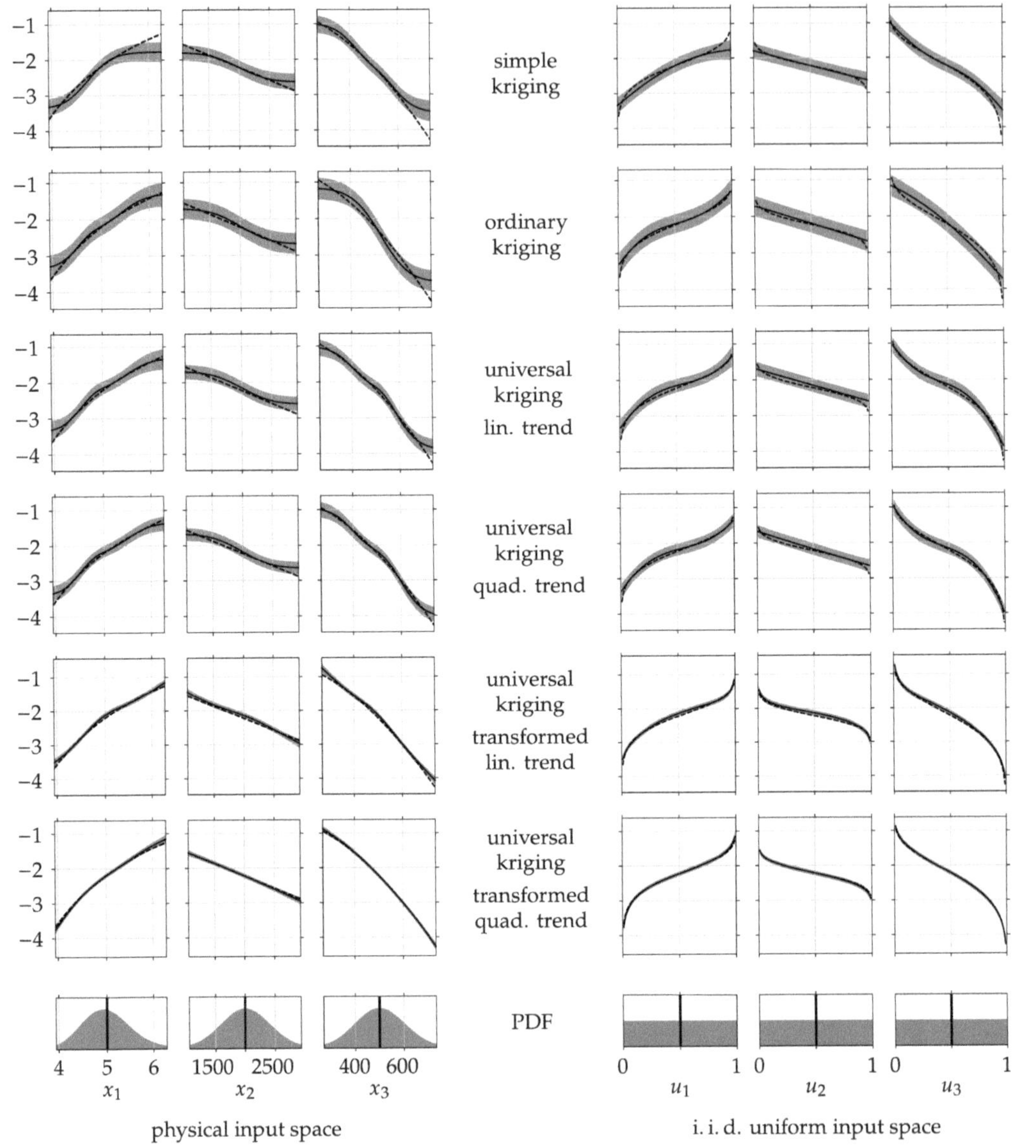


**Figure 3.3:** Visualization of benchmark problem #4 (short column function): True function (dashed line) and GPR mean (solid line) with variance (grey shaded area) for the investigated GPR methods (rows) and input parameters (columns), shown in the physical (left) and i. i. d. uniform (right) input spaces. The plots represent cuts through the hypersurface where one input parameter varies while all others remain fixed at their mean values (shown in PDF on the bottom).

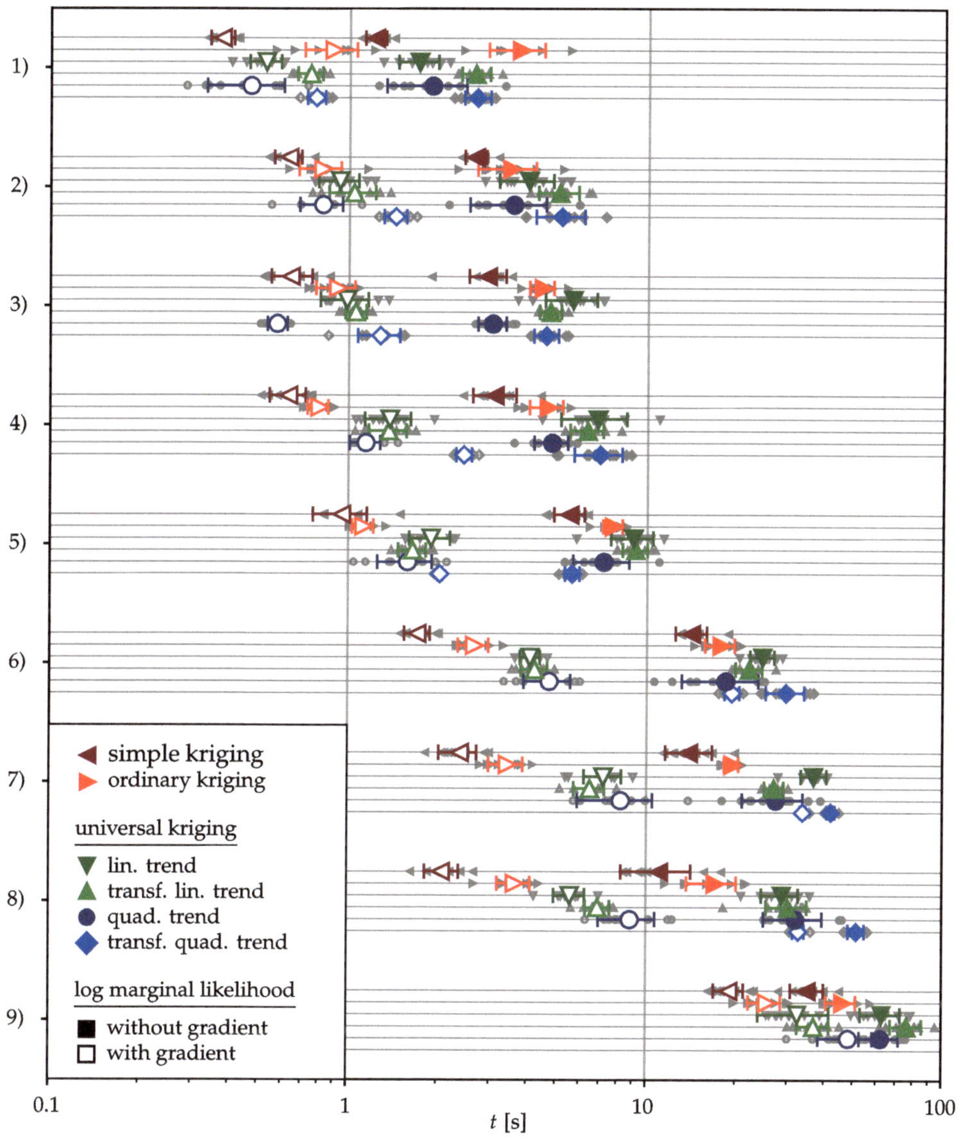


**Figure 3.4:** Computation times for constructing surrogate models across all benchmark problems and investigated GPR methods. Results are shown for cases with (empty markers) and without (filled markers) the incorporation of the gradient of the log marginal likelihood in hyperparameter optimization. Mean values and standard deviations across 10 experiments are presented for each case.

## 3.2 Expert-informed multi-objective optimization concepts

In many studies, parameter optimization involves multiple objective functions. While optimizing a scalar-valued single-objective function

$$\mathbf{u}_{\text{opt}} = \underset{\mathbf{u}\in\mathcal{S}}{\text{argmin}}\{f(\mathbf{u})\}$$

can often be achieved efficiently using various local or global optimization methods, the task becomes more complex and less straightforward when optimizing multiple objective functions

$$\mathbf{u}_{\text{opt}} = \underset{\mathbf{u}\in\mathcal{S}}{\text{argmin}}\{f_j(\mathbf{u}), j = 1 \ldots J\} \,.$$

To address this challenge, multi-objective optimization methods have been developed, typically aiming to identify Pareto-optimal fronts – sets of non-dominated solutions where an improvement in one objective can only be achieved at the expense of at least one other objective [29].

Due to the high dimensionality of many practical problems, it is often beneficial to assign weights to individual objectives and combine them into a single objective function using p-norm scalarization

$$f(\mathbf{u}) = \left( \sum_{j=1}^{J} w_j |f_j(\mathbf{u})|^p \right)^{1/p} ,$$

where $p = 1$ corresponds to a weighted sum, $p = 2$ to a squared sum, and $p = \infty$ to the Tchebycheff norm. However, a fixed set of weights does not provide a comprehensive understanding of the optimal parameters, as these may be highly sensitive to the choice of weights. Therefore, it is essential to explore variations in the weight selection. To address this, two approaches are proposed: (a) introducing uncertainty in the weights to analyze the sensitivity (spread) in the optimal parameters, and (2) independently varying individual weights to assess the isolated influence of each objective. Both methods rely on a default weight combination $\mathbf{w}_0 = (w_{0,j}, j = 1 \ldots J)$, determined based on expert judgment. Without loss of generality, the weights $\mathbf{w}_0$ are scaled such that their sum satisfies $\sum_{j=1}^{J} w_{0,j} = 1$, allowing them to be interpreted as the relative importance of the objectives in determining the optimal parameters.

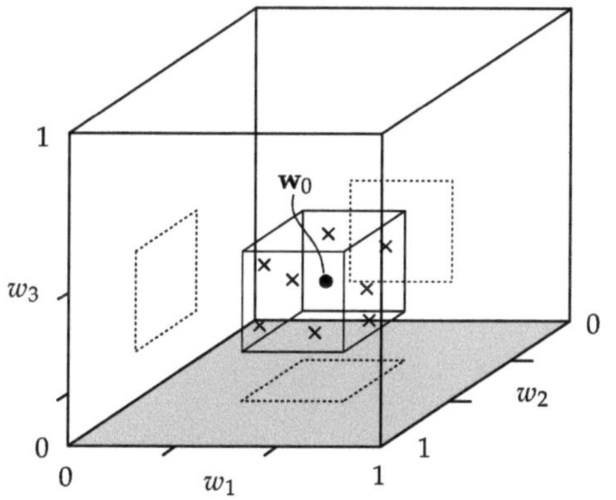


a) *weight uncertainty* strategy

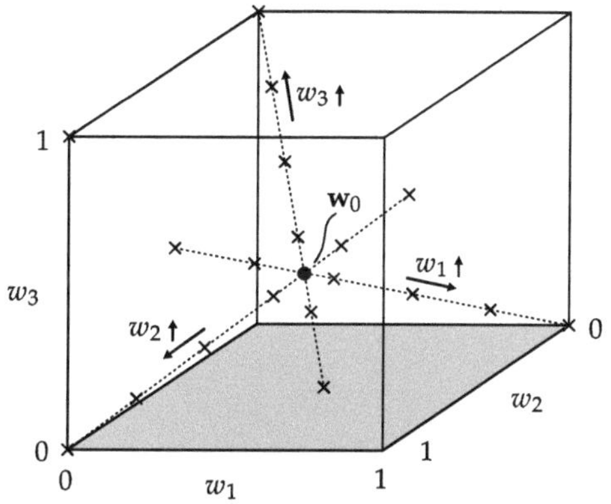


b) *weight variation* strategy

**Figure 3.5:** Visualization of multi-objective optimization strategies. Solid point $\mathbf{w}_0$ corresponds to the default weight combination, crosses correspond to samples for weight combinations used for separate optimizations.

(a) *Weight uncertainty*
To assess the sensitivity of optimal parameters to variations in objective weights, uncertainty is introduced into the weights. Each weight is sampled from a uniform interval, as illustrated for a three-dimensional case in Fig. 3.5a. Within this hyper-rectangle, a MC simulation is conducted, where each sample represents a distinct single-objective optimization problem with a specific weight combination. The resulting distributions of optimal parameters and their corresponding objective values can then be visualized using histograms.

(b) *Weight variation*
To assess the dependence of optimal parameters on individual objective weights, the weight of one objective is systematically varied step-wise between 0 and 1, while all other weights are linearly scaled to ensure that $\sum_{j=1}^{J} w_j = 1$ remains valid. This approach preserves the relative proportions among the remaining weights, isolating the effect of varying a single weight. The process is repeated for all weights, resulting in distinct lines of points in the weight space, as illustrated in Fig. 3.5b.

These two methods offer complementary approaches to analyzing how objective weights influence optimal model parameters. By assessing the sensitivity of these parameters to variations in the weights, the results can reveal whether the parameters are robust to such changes and quantify their values. Traditional validation is not applicable in this context, as the methods do not produce objectively verifiable results but instead provide a structured framework for exploring the problem.

# 4 Uncertainty quantification and parameter optimization in simulations of the West African monsoon

Simulating the meteorological system of the WAM is associated with substantial uncertainties. This section introduces the meteorological model used in this study, specifically the ICON model developed by the DWD, along with the selected model parameters. The applied meteorological data are then described in detail. Building on the previous chapter, surrogate modeling [160] and optimization [161] techniques are employed to assess the uncertainty contributions of meteorological parameters and identify opportunities for model improvement.

## 4.1 Meteorological model

The ICON model is a global numerical weather prediction model developed by the DWD in collaboration with the Max Planck Institute for Meteorology [152]. It employs a non-hydrostatic formulation on an icosahedral grid to simulate atmospheric processes across multiple scales. This section introduces the specific ICON model configuration used in this study. Additionally, the ICON model parameters considered for surrogate modeling and optimization are described, along with the assigned PDFs that represent their epistemic uncertainties.

### 4.1.1 ICON model configuration

The ICON model, the operational forecast system of the DWD, is used as the full-physics numerical model to simulate the WAM. For this purpose, version 2.5.0 is employed in a limited-area nested configuration, with a 26 km grid spacing for the outer region and a 13 km grid spacing for the inner region. The outer domain extends from 28°W to 34°E

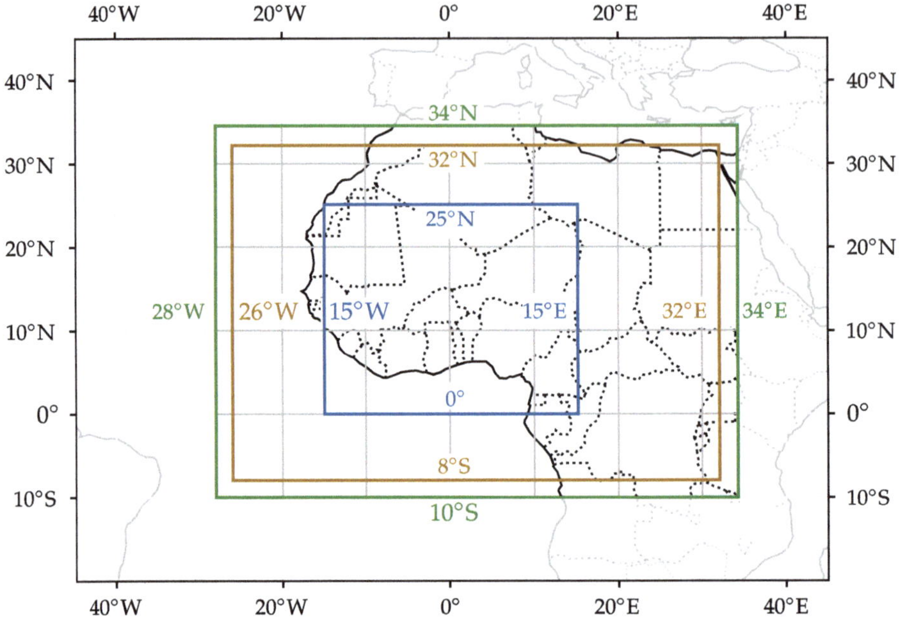


**Figure 4.1:** ICON model configuration, outer domain with 26km grid spacing (green), inner domain with 13km grid spacing (brown) and domain for which simulation output data are stored (blue).

and 10°S to 34°N, while the nested domain is 2° (~ 220 km) smaller in each direction (Fig. 4.1). At the outer boundary, ERA5 reanalysis data [49], provided by the European Centre for Medium-Range Weather Forecasts (ECMWF), are used. Although ERA5 data are available hourly, updates in the simulations are applied every six hours to limit data volume and computational cost. Apart from this, the model setup, including all namelist parameters, follows the DWD's operational global configuration. Previous studies, such as Pante and Knippertz [108], demonstrated reasonable simulation results for the West African region using a similar model setup, though convection parameterization posed challenges for precipitation forecasting. To isolate model parameter sensitivities from those induced by weather noise and to minimize the influence of initial conditions, a sufficiently long simulation period is required. At the same time, the focus remains on capturing the peak of the WAM during boreal summer. To balance these aspects, August data from 2016 to 2019 are considered. Each simulation starts on July 21st, runs for 41 days, and only data from August 1st to August 31st are analyzed to reduce the impact of initial conditions. Preliminary tests using simulations from a single year showed high fluctuations in the considered QoIs and low surrogate model accuracy, primarily due to aleatoric uncertainties arising from the atmosphere's small-scale chaotic behavior. To obtain a more robust estimate while maintaining manageable computational cost, analyzing the rainy season across four years provided a reasonable compromise. The meteorological variables from the model output are therefore averaged over these four August periods to represent a multi-year mean state and serve as training points for the surrogate models. Using data from four different years also captures variations in SSTs, which are prescribed as boundary conditions based on SST analyses at model initialization. During the simulation, SSTs are incrementally updated according to their annual climatological cycle [116].

### 4.1.2 Model parameter selection

Within the ICON ensemble prediction system, forecast uncertainties are assessed by perturbing both initial conditions and physical model parameters. The ensemble uses a local ensemble transform Kalman filter [53] for initial condition perturbations and applies randomized variations to predefined physical tuning parameters to account for model uncertainties. Since these physical parameters already provide an established basis for parametric uncertainty sources, this study focuses on six parameters that span a broad range of the model's physics and are expected to have a significant impact on the WAM region. Their selection was guided by expert judgment from the Institute of Meteorology and Climate Research – Troposphere Research at the Karlsruhe Institute of Technology. These parameters include the grid-scale microphysics (*zvz0i*), turbulence (*tkhmin*), land-surface interaction (*c_soil*) and the parameterization of deep convection (*entrorg, rhebc_land_trop, rcucov_trop*). For the purpose of the analysis in this work, the parameters are grouped into three pairs with regard to their physical implication, namely deep-cloud (*entrorg, zvz0i*), below-cloud (*rhebc_land_trop, rcucov_trop*) and boundary-layer (*tkhmin, c_soil*) parameters (see Table 4.1).

The entrainment rate (*entrorg*) controls the mixing of ambient air into convective plumes. Depending on the free-tropospheric humidity, higher *entrorg* values may reduce buoyancy within the convective plumes, potentially leading to decreased convective rainfall. The terminal fall speed of ice crystals (*zvz0i*) influences the lifetime of cirrus clouds and, consequently, the average high-level cloud cover. Particularly in the tropics, this parameter can strongly affect cloud-radiative heating rates, which in turn may impact large-scale atmospheric circulation. Despite their distinct physical roles, the entrainment rate and terminal fall velocity of ice exhibit similar overall effects: Reinert et al. [116] found that reduced entrainment increases the altitude of tropical convection and enhances the production of cloud ice in the upper troposphere. To maintain radiative forcing at a similar level, this effect must be counterbalanced by faster cloud ice sedimentation. For this reason, the DWD varies these two parameters inversely in operational ensemble physics perturbations [116]. Below-cloud parameters primarily influence evaporation processes in convective regions. The parameter *rhebc_land_trop*

**Table 4.1:** Selected uncertain model parameters including a brief description, the assumed PDF, and the corresponding physical unit.

| | model parameter | description | PDF[1] | unit |
|---|---|---|---|---|
| a) | entrorg | entrainment parameter valid for dx=20 km (depends on model resolution) | $\mathcal{LN}(\mu = -6.3,\ \sigma = 0.18)$ | $m^{-1}$ |
| b) | zvz0i | terminal fall velocity of ice | $\mathcal{LN}(\mu = 0.22,\ \sigma = 0.40)$ | $m\,s^{-1}$ |
| c) | rhebc_land_trop | relative humidity threshold for onset of evaporation below cloud base over land in the tropics | $\mathcal{B}(\alpha = 30,\ \beta = 10)$ | – |
| d) | rcucov_trop | convective area fraction used for computing evaporation below cloud base in the tropics | $\mathcal{LN}(\mu = -3.0,\ \sigma = 0.27)$ | – |
| e) | tkhmin | scaling factor for minimum vertical diffusion coefficient for heat and moisture | $\mathcal{LN}(\mu = -0.29,\ \sigma = 0.36)$ | $m^2\,s^{-1}$ |
| f) | c_soil | surface area density of the (evaporative) soil surface | $\mathcal{N}(\mu = 1.0,\ \sigma = 0.34)$ | – |

[1] $\mathcal{LN}$: log-normal distribution ($\mu$ and $\sigma$ are the mean and standard deviation of the variable's natural logarithm), $\mathcal{N}$: normal distribution, $\mathcal{B}$: beta-distribution.

defines a relative humidity threshold below which sub-cloud evaporation occurs in convectively active grid cells over tropical land areas. The parameter *rcucov_trop* represents the areal fraction of convection within a grid cell used for computing evaporation below cloud base. While *rhebc_land_trop* affects regions where relative humidity is near its threshold value, *rcucov_trop* has a broader impact, influencing evaporation across most of the domain. The turbulent diffusion coefficient (*tkhmin*) regulates the vertical mixing of heat and moisture, which can influence cloud formation. In reality, some level of vertical diffusion is always present. However, under highly stable conditions with weak vertical wind shear, turbulence parameterizations tend to underestimate this diffusion. To compensate for this, a minimum diffusion value (*tkhmin*) is imposed in the model [112]. The parameter *c_soil* represents the fraction of soil contributing to evaporation, expressed as a unitless fraction. Higher values increase near-surface humidity and may enhance cloud cover formation. Particularly for *entrorg*, *rhebc_land_trop* and *rcucov_trop* the net effect on area- and time-integrated precipitation is uncertain, as it strongly depends on the meteorological context.

In meteorological studies and operational forecasts, model parameters are often assumed to follow uniform distributions over an estimated range of plausible values [141]. This assumption is also applied in the operational ensemble forecasts of the DWD [116]. The use of uniform distributions is reasonable when limited prior information is available, as the primary objective is to introduce variability into ensemble forecasts and ensure that the forecast uncertainty adequately reflects the inherent limitations of predictability. While uniform distributions are a practical choice for ensemble forecasting, they are less suitable for GSA. A uniform distribution imposes an artificial discontinuity in the probability density at its boundaries, meaning that values near the range limits disproportionately influence GSA results, while values just outside the range contribute nothing. Consequently, alternative, non-uniform PDFs provide a more appropriate representation of parameter uncertainty.

Although defining such PDFs presents a challenge in its own right, they are considered more appropriate for this study. Non-uniform PDFs for the parameters analyzed here have already been applied by Lang et al. [66] and Ollinaho et al. [105], where normal and log-normal distributions were used to represent parameter uncertainties. The primary source for defining the PDFs in this study is the parameter ranges and mean values used in operational ensemble forecasts by the DWD [15]. These are supplemented by further expert knowledge. The probability distributions are chosen to ensure consistency with physical constraints and symmetry considerations. Parameters that must remain strictly positive are described by functions that only take positive values, such as log-normal PDFs. Parameters bounded between 0 and 1, representing fractional values, are better described by beta distributions. In the case of the parameter *c_soil*, the standard deviation of the normal distribution is low enough to ensure positive values and is used to assume a symmetric distribution for simplicity. The selected model parameters and their assigned PDFs are presented in Table 4.1, while visualizations of the distributions are provided in the results section (Fig. 4.3, at the bottom).

In an operational forecasting framework, where the primary objective is to improve predictive accuracy, temporally and spatially varying parameter values may enhance flexibility and overall forecast performance. However, this study focuses on identifying systematic relationships between parameter values and simulation outcomes. For the sake of simplicity and interpretability, model parameters are kept constant in both time and space within individual simulations. Consequently, each training point corresponds to a fixed set of six model parameters, which are used in a single ICON model simulation.

As described in Section 2.1.4, training points, i. e. model parameter configurations, are generated using maximin Latin Hypercube Sampling (LHS) to ensure a well-distributed sampling across the parameter space. A total of $n = 60$ training points are selected in the $p = 6$ dimensional input space. The required number of training points strongly depends on the nonlinearity of the investigated system. However, following the heuristic $n = 10p$ proposed by Loeppky, Sacks, and Welch [76], this choice is adopted while recognizing the necessity of model validation. As outlined in Section 3.1.1, the sampling is initially performed in a unit hypercube, after which the inverse Rosenblatt transformation $\mathbf{x_i} = \mathcal{T}_{\text{ros}}^{-1}(\mathbf{u_i})\,(i = 1 \ldots n)$ is applied to map the sampled points to the physical parameter space used in the ICON model simulations. Since the six investigated model parameters are assumed to be independent, the transformation simplifies to the inverse CDF for each parameter $\mathbf{x_i} = \mathcal{T}_{\text{ros}}^{-1}(\mathbf{u}_i) = (\text{CDF}_j^{-1}(u_{ij}), j = 1 \ldots 6)^\top$.

## 4.2 Meteorological data

The ICON model computes a wide range of meteorological variables, including state and diagnostic variables, which can be stored as simulation output depending on a chosen spatial and temporal discretization. The selection of stored data, both from ICON simulations and reference datasets used for comparison, must balance its utility with storage constraints. This section outlines the meteorological field data directly retained for each ICON model simulation corresponding to each parameter configuration $\mathbf{x_i}\,(i = 1 \ldots n)$, as well as from reference data sets, requiring approximately 475 gigabytes of storage in total. Additionally, to gain deeper insights into specific characteristics of the WAM, scalar QoIs are derived from both ICON output data and reference data, as described in the following.

### 4.2.1 Meteorological fields

Simulation and reference data are stored at a horizontal resolution of 0.1° over the region spanning 0°N to 25°N and 15°W to 15°E (see Fig. 4.1) for the period August 1–31

**Table 4.2:** Selected meteorological fields.

| j | meteorological field | unit |
|---|---|---|
| 1 | accumulated precipitation | mm per month |
| 2 | 2 m temperature | K |
| 3 | 2 m dew point temperature | K |
| 4 | mean sea-level pressure (MSLP) | hPa |
| 5 | column-integrated water vapor | $kg\,m^{-2}$ |
| 6 | high-level cloud cover | % |
| 7 | mid-level cloud cover | % |
| 8 | low-level cloud cover | % |
| 9 | u-wind at 600 hPa | $m\,s^{-1}$ |
| 10 | u-wind at 200 hPa | $m\,s^{-1}$ |
| 11 | v-wind at 600 hPa | $m\,s^{-1}$ |
| 12 | v-wind at 200 hPa | $m\,s^{-1}$ |

in the years 2016, 2017, 2018 and 2019. The selected model outputs, which represent key characteristics of the WAM, are listed below, along with their respective temporal resolutions. Notably, a finer temporal resolution is applied to cloud cover data to better capture its anticipated higher variability.

The ICON model output from all 60 model parameter configurations is stored for all meteorological field variables at the specified spatial and temporal resolutions. As reference data, the Global Precipitation Measurement (GPM) Integrated Multi-satellitE Retrievals (IMERG) data [52] are used for precipitation due to its high spatial and temporal resolution, which are essential for accurately capturing the variability of rainfall in the WAM region. For other atmospheric variables, such as temperature, pressure, and wind patterns, the ERA5 reanalysis data are utilized. To facilitate comparison across datasets, the ICON model output (native resolution of 0.1°), ERA5 data (native resolution of 0.25°), and GPM IMERG data (native resolution of 0.1°) are linearly remapped onto a common rectangular grid with a mesh size of 0.5°. The data are averaged over the four August months. The spatial resolution is chosen as a compromise between accuracy and computational efficiency, particularly for the parameter optimization process.

### 4.2.2 Quantities of Interest

Using the meteorological fields, specific characteristics of the WAM system can be derived to analyze the influence of model parameters in greater detail. In this study, the QoIs describing these characteristics are defined as scalar quantities. The field data has been processed such that only the temporal average over August across all four years (2016-2019) is considered, with a spatial resolution of 0.5°. The selected QoIs and a brief description of their computation are provided in Table 4.3.

In meteorological studies, the computation of such quantities often involves subjective decisions, depending on the available data and the specific objectives of the analysis. For some of the QoIs considered here, such decisions were necessary and will be briefly discussed in the following sections.

The consideration of latitudinal values for several features is only meaningful within certain longitudinal ranges, as outside these ranges, the features may be absent or influenced by factors that are not relevant to this study. Including such regions could obscure the influence of model parameters on QoIs. The precipitation latitude provides a measure of the north-south shift of the rainbelt. The longitudinal range for this computation is selected between 12°W and 2°E to minimize the impact of distinct topographical features, such as the Guinea Highlands to the west and the Cameroon Line and wet Niger Delta to the east. The ITD marks the interface where dry northeasterly winds from the Sahara meet moist southwesterly winds from the tropical Atlantic Ocean. While its latitude could be determined from 3D wind fields at different altitudes, the available meteorological fields (Table 4.2), which were selected due to data storage constraints, necessitate an alternative approach. Therefore, the ITD latitude is computed using the 2 m dew point temperature, which effectively captures the sharp transition in moisture content near the surface. Specifically, the latitude corresponding to a 2 m dew point temperature of 14 °C is used [24]. The computation is restricted to a longitudinal range of 12°W to 8°E, which is broader than that for the precipitation center latitude, allowing for a more robust analysis, as the ITD remains relatively stable over a wider longitudinal extent. For the southern boundary of the SHL, computed based on a mean sea-level pressure (MSLP) threshold of 1009 hPa, a longitudinal range of 9°W to 1°W is chosen, as the SHL is primarily confined to this region. Expanding the latitudinal range or increasing the pressure threshold could result in cases where the threshold is no longer met at certain longitudes, making the characterization less meaningful. The southern boundary is used instead of the center latitude since, given the study region's northern limit at 25°N, the center latitude cannot be robustly determined. In potential future studies, it may be beneficial to extend

**Table 4.3:** Selected QoIs derived from meteorological fields.

| j | QoI | unit | description |
|---|---|---|---|
| 1 | accumulated precipitation average | mm per month | averaged accumulated precipitation[1] |
| 2 | precipitation center latitude | °N | weighted average of latitudes between 12°W and 2°E, using accumulated precipitation as weighting factor |
| 3 | 2 m temperature average | K | averaged 2 m temperature[1] |
| 4 | 2 m dew point temperature average | K | averaged 2 m dew point temperature[1] |
| 5 | ITD latitude | °N | average of latitudes corresponding to a 14 °C 2 m dew point temperature for all longitudes from 12°W to 8°E |
| 6 | SHL pressure | hPa | average of the 10 % lowest MSLP values within the region from 15°N to 25°N and 15°W to 5°E |
| 7 | SHL southern boundary | °N | average of latitudes corresponding to a 1009 hPa MSLP threshold for all longitudes from 9°W to 1°W |
| 8 | column-integrated water vapor average | $kg\,m^{-2}$ | averaged column-integrated water vapor[1] |
| 9 | high-level cloud cover average | % | averaged high-level cloud cover[1] |
| 10 | mid-level cloud cover average | % | averaged mid-level cloud cover[1] |
| 11 | low-level cloud cover average | % | averaged low-level cloud cover[1] |
| 12 | AEJ speed | $m\,s^{-1}$ | averaged u-winds at 600 hPa along the AEJ latitude (13) |
| 13 | AEJ latitude | °N | weighted average of latitudes, using u-winds at 600 hPa exponentiated by 3 as weighting factor |
| 14 | TEJ speed | $m\,s^{-1}$ | averaged u-winds at 200 hPa along the TEJ latitude (15) |
| 15 | TEJ latitude | °N | weighted average of latitudes, using u-winds at 200 hPa exponentiated by 3 as weighting factor |

[1] average value over all grid points of the study domain (15°W to 15°E, 0°N to 25°N)

the region where output data is stored further north to enable a more comprehensive analysis of the SHL.

The SHL, one of the main drivers of the WAM, is characterized by its strength. To quantify this, the average pressure field is computed for each August month within the region 15°N to 25°N and 15°W to 5°E, where the heat low is expected based on climatological studies [68]. For a robust characterization, the MSLP is averaged over the 10 % lowest values within this region, rather than considering all grid points or selecting only the absolute minimum MSLP value. Averaging all values would obscure the localized pressure minimum, while selecting only a single grid point would be overly sensitive to fluctuations due to the chaotic nature of the system. This approach ensures a balance between stability and representativeness in capturing the intensity of the heat low.

For the computation of jet latitudes, using only the latitude of the maximum zonal wind speed proved to be a non-robust measure, as it is highly sensitive to small parameter variations and prone to fluctuations due to the chaotic nature of the atmosphere. To improve robustness, neighboring latitudes are also taken into account. First, the averaged zonal wind speed for each latitude on the grid is computed. However, the resulting distribution remains relatively flat, making it difficult to reliably determine the latitude of maximum wind speed. To address this issue, the average wind values are exponentiated, which enhances the sharpness of the profile and reduces the influence of relatively high wind values located far from the jet center. In this study, an exponent of 3 was found to yield meaningful results. Finally, the weighted average of latitudes is computed using the exponentiated wind values as weights. Without this exponentiation strategy, the chosen latitudinal range for analysis would significantly affect the result, which should be avoided. When computing the jet speeds based on the obtained latitudes, it is important to note that the maximum instantaneous jet speed is considerably higher, as only the time-averaged wind speeds are considered for the sake of enhanced robustness.

## 4.3 Surrogate models for meteorological variables

In this section, surrogate models are developed to characterize the influence of the considered model parameters on WAM characteristics. PCR is employed to construct surrogate models for meteorological fields, while GPR is used for scalar QoIs. The surrogate models for meteorological fields provide insights into how parameter variations affect the geographical distribution of key meteorological variables. In contrast, using scalar QoIs sacrifices spatial variability in favor of a more robust analysis, emphasizing specific aspects of the WAM system. All surrogate models are trained in the i. i. d. uniform input space based on the $n = 60$ training points (Section 4.1.2).

### 4.3.1 Meteorological fields

The meteorological fields for the 60 training points are used to construct surrogate models via PCR to describe the relationship between the six model parameters and the meteorological fields. The temporally averaged meteorological field for variable $j\,(j = 1\dots12)$ at training point $\mathbf{u}_i(i = 1\dots n)$ is given by $\{\mathbf{M}_i : M_{ij}^{kl}, k = 1\dots50, l = 1\dots60\}$ where 50 latitudinal and 60 longitudinal grid points are considered. Following the surrogate modeling procedure outlined in Section 2.3, the mean $\mu_j^{kl}$ (Eq. 2.10) and standard deviation $\sigma_j^{kl}$ (Eq. 2.11) for each meteorological variable $j\,(j = 1\dots12)$ are computed. The resulting surrogate model for the meteorological field, as formulated in Eq. 2.12, can be represented as

$$\hat{M}_j^{kl}(\mathbf{u}) \approx \mu_j^{kl} + \sigma^{kl}\tilde{M}_j^{kl}(\mathbf{u}) \quad \text{with} \quad \tilde{M}_j^{kl}(\mathbf{u}) = \sum_{m=1}^{P} C_{jm}(\mathbf{u})P_{jm}^{kl} \,.$$

The coefficients are modeled using a linear trend with respect to the physical model parameters, as described in Eq. 3.3. In this study, selecting $P = 3$ principal fields provides high accuracy across all meteorological variables, i. e. the changes in NMSE errors were less than 1 % when including additional principal fields, while representing a good compromise between accuracy and computational efficiency.

### 4.3.2 Quantities of Interest

The QoIs derived from the meteorological fields at the 60 training points are used to construct surrogate models via GPR, capturing the relationship between the six model parameters and the scalar QoIs. To enhance numerical stability and ensure the well-posedness of the problem, as is well established, the QoIs are standardized for the surrogate modeling process [55], ensuring zero mean and unit variance. The value of QoI $j\,(j = 1\dots15)$ for training point $\mathbf{u}_i\,(i = 1\dots n)$ is given by $y_{\text{orig},ij}$. The standardized data are then computed as

$$y_{ij} = \frac{1}{\sigma_{\text{QoI}_j}}(y_{\text{orig},ij} - \mu_{\text{QoI}_j})\,, \tag{4.1}$$

with the mean and standard deviation defined by

$$\mu_{\mathrm{QoI}_j} = \frac{1}{n}\sum_{i=1}^{n} y_{\mathrm{orig},ij}\,, \tag{4.2}$$

$$\sigma_{\mathrm{QoI}_j} = \sqrt{\frac{1}{n}\sum_{i=1}^{n}\left(y_{\mathrm{orig},ij} - \mu_{\mathrm{QoI}_j}\right)^2}\,. \tag{4.3}$$

Following the surrogate modeling procedure described in Section 2.2.2, the predictive mean and predictive covariance for prediction points $\mathbf{U}_\star$ of a given QoI $j$ are

$$\hat{\mathbf{y}}_{\mathrm{QoI},j}(\mathbf{U}_\star) = \mathbf{K}_\star^\top \mathbf{K}^{-1}\mathbf{y}_j + \mathbf{R}^\top \bar{\boldsymbol{\beta}}$$
$$\hat{\sigma}_{\mathrm{QoI},j}(\mathbf{U}_\star) = \mathbf{K}_{\star\star} - \mathbf{K}_\star^\top \mathbf{K}^{-1}\mathbf{K}_\star + \mathbf{R}^\top(\mathbf{H}\mathbf{K}^{-1}\mathbf{H}^\top)^{-1}\mathbf{R}$$

with $\mathbf{R} = \mathbf{H}_\star - \mathbf{H}\mathbf{K}^{-1}\mathbf{K}_\star$ and $\bar{\boldsymbol{\beta}} = (\mathbf{H}\mathbf{K}^{-1}\mathbf{H}^\top)^{-1}\mathbf{H}\mathbf{K}^{-1}\mathbf{y}_j$.

For the trend functions $\mathbf{h}(\mathbf{u})$ (Eq. 2.5 and Eq. 2.6), a linear trend w. r. t. the physical model parameters $\mathbf{x}$ is used as elaborated in Section 3.1.2. Given $\mu_{\mathrm{QoI}_j}$ (Eq. 4.2) and $\sigma_{\mathrm{QoI}_j}$ (Eq. 4.3), the surrogate model results for each individual QoI $j$ can be transformed back to the original data space for visualization and analysis.

### 4.3.3 Validation

In this study, only temporally averaged quantities over August from 2016 to 2019 are considered. Due to the complex and chaotic nature of the atmosphere, long-term predictions of meteorological variables in the tropics are particularly challenging, often resulting in larger deviations between ICON model simulations and reference data. However, the focus here is on the climatological mean state for the given period, where simulation data is expected to better align with reference data. Validation is therefore conducted only for temporally averaged meteorological fields and the QoIs derived from them. First, the ICON simulation results are compared against reference data. While these results depend on the model parameters, the simulation output – especially when using default parameters – should reasonably match the reference data, confirming the plausibility of the model configuration and computation strategies for QoIs. Nevertheless, optimizing the model parameters to improve this agreement remains a key objective of this study. Next, the surrogate models for the meteorological fields and QoIs are validated. Given the computational demands of weather models, no separate validation dataset is used. Instead, leave-$k$-out cross-validation is employed to assess the performance of the surrogate models.

For validation of the mean ICON model output fields against reference data, the RMSE and NMSE from Section 2.4 are adapted as

$$\mathrm{RMSE}_j = \sqrt{\frac{1}{KL}\sum_{k=1}^{K}\sum_{l=1}^{L}(\mu_j^{kl} - \hat{M}_{\mathrm{ref},j}^{kl})^2}\,,$$

$$\mathrm{NMSE}_j = \frac{1}{\sigma_{Mj}^2}\frac{1}{KL}\sum_{k=1}^{K}\sum_{l=1}^{L}(\mu_j^{kl} - \hat{M}_{\mathrm{ref},j}^{kl})^2\,,$$

where $\hat{M}_{\mathrm{ref},j}^{kl}$ represents the reanalysis data for meteorological variable $j$ at grid points $(k,l)$ (see 4.2.1 for details). The validation results of the ICON model output are presented in Table 4.4. The RMSE for accumulated precipitation is 47.7 mm, which corresponds to 15.8 % in NMSE. An inspection of the spatial distribution shows that the differences are mostly due to wetter conditions along the rainy southwestern coast of West Africa and Niger Delta region in ERA5. Differences in cloud cover are also substantial. While high clouds exhibit the best agreement (7.6 % RMSE), low- and mid-level cloud cover is substantially higher in ICON with RMSEs of 15.5 % and 6 %, respectively. These values correspond to NMSEs of 46.9 % and 24.1 %, indicating notable discrepancies. Low clouds over tropical West Africa are controlled by a subtle balance of advective, radiative and turbulent processes [77], and differences between models tend to be large [45]. Given that cloud cover and precipitation are strongly influenced by model parameterizations, such differences are expected. Moreover, while ERA5 has undergone significant improvements compared to earlier reanalysis products, it may still have uncertainties in these quantities [33]. The other moisture-related variables, column-integrated water vapor and 2 m dew point temperature, show only minor deviations, as does MSLP. However, differences in 2 m temperature are more pronounced with an RMSE of 1.7 K and a NMSE of 20.1 %. This discrepancy is primarily due to warmer temperatures over the Sahara in ICON (not shown). Modeling near-surface temperature in desert regions is particularly challenging due to intense solar heating and turbulent surface sensible heat fluxes, which can lead to superadiabatic lapse rates in the lowest atmospheric layers [65]. Finally, the four wind variables exhibit good agreement with the exception of v-winds at 600 hPa, where the NMSE is 26.4 %, corresponding to a RMSE of only 0.6 $\mathrm{m\,s^{-1}}$. This deviation is primarily attributed to stronger northerlies over the Sahara in ERA5 (not shown). These validation results confirm that the model setup is generally valid, despite notable discrepancies in certain quantities. Further reducing these discrepancies will be a key objective of the parameter optimization studies in Section 4.5, where the differences between ICON and reference data will be examined in greater detail, with a particular focus on their physical implications.

The accuracy of the obtained surrogate models for meteorological fields and QoIs depends on several factors, including the number of training points, the modeling methodology, the choice of ansatz functions (e. g., coefficient ansatz, number of principal fields for PCR, or kernel functions for GPR), and the presence of nonlinearities or chaotic behavior within the physical model. As a compromise between accuracy and computation efficiency, leave-$k$-out cross-validation with $k = 2$ is applied individually to all meteorological fields (Eq. 2.18,2.19) and QoIs (Eq. 2.16,2.17). The validation results are presented in Table 4.5. The observed errors in the validation process include both aleatoric uncertainties inherent in weather simulations (which are inevitable due to the chaotic nature of the system) and uncertainties introduced by the surrogate models. It is important to acknowledge that large errors do not necessarily indicate low surrogate model accuracy, as they may also reflect substantial aleatoric uncertainties in the respective quantities. A small RMSE (or NMSE) suggests that both surrogate model accuracy is high and aleatoric uncertainty is low, providing confidence for the subsequent analyses. Regarding the validation of meteorological fields, the presence of regional variability means that comparing grid point data within the validation procedure inherently results in substantially larger errors compared to domain-averaged comparisons used for validating QoIs. Small NMSE values are evident across all meteorological fields, except for high-level cloud cover (NMSE 8.06 %) and v-winds at 600 hPa (NMSE 4.37 %). For GPR, the predictive variance (Eq. 2.9) can be considered as a measure for the uncertainty of the surrogate model. The noise variance $\sigma_2$ of the kernel function (Eq. 2.1) as a result of maximum likelihood estimation provides an estimate of the aleatoric uncertainty (see Table 4.5). Since the values of $\sigma_2$ are generally lower but of a similar magnitude compared to the RMSE, this suggests that a significant proportion of the observed validation errors may be attributed to aleatoric uncertainties inherent in the weather model. NMSE values are considered small for all QoIs except for the AEJ speed and precipitation latitude. However, the small RMSEs for the v-winds at 600 hPa, AEJ speed and precipitation latitude indicate that the absolute errors are very small. The variance $\sigma_y^2$ (Eq. 2.17, corresponding to $\sigma_{\mathrm{QoI}_j}$ in Eq. 4.1) used for normalization is also small due to low absolute variations in these variables, leading to larger NMSE values. Consequently, high NMSE values in these cases do not affect the overall validity of this study. Furthermore, these variables will be assigned low weights in the optimization process, so their larger NMSE values are not expected to impact the study's overall validity. Errors could be further reduced by increasing the number of training points or averaging over a longer period (i. e. incorporating more years of data).

**Table 4.4:** Validation results of the ICON model output compared to reference data.

| j | meteorological field | $\mathrm{RMSE}_j$ ($\mathrm{NMSE}_j$) |
|---|---|---|
| 1 | accumulated precipitation | 47.7 mm mo$^{-1}$ (15.78%) |
| 2 | 2 m temperature | 1.70 K (20.09%) |
| 3 | 2 m dew point temperature | 0.964 K (3.46%) |
| 4 | MSLP | 50.7 Pa (4.76%) |
| 5 | column-integrated water vapor | 1.44 kg m$^{-2}$ (2.58%) |
| 6 | high-level cloud cover | 7.60 % (8.80%) |
| 7 | mid-level cloud cover | 6.03 % (24.14%) |
| 8 | low-level cloud cover | 15.5 % (46.92%) |
| 9 | u-wind at 600 hPa | 0.609 m s$^{-1}$ (3.34%) |
| 10 | u-wind at 200 hPa | 1.37 m s$^{-1}$ (4.19%) |
| 11 | v-wind at 600 hPa | 0.599 m s$^{-1}$ (26.43%) |
| 12 | v-wind at 200 hPa | 0.598 m s$^{-1}$ (3.82%) |

**Table 4.5:** Validation results for surrogate models of meteorological fields and QoIs. For QoIs, $\sigma_2$ represents the noise term in the kernel function (see Eq. 2.1), obtained from hyperparameter estimation.

| j | meteorological field [unit] | $RMSE_j$ ($NMSE_j$) | | $\sigma_2$ | QoI [unit] | j |
|---|---|---|---|---|---|---|
| 1 | accumulated precipitation [mm mo$^{-1}$] | 15.5 (1.56%) | 1.365 (11.64%) | 0.385 | average [mm mo$^{-1}$] | 1 |
| | | | 0.057 (28.98%) | 0.049 | center latitude [°N] | 2 |
| 2 | 2 m temperature [K] | 0.300 (0.45%) | 0.054 (3.88%) | 0.025 | average [K] | 3 |
| 3 | 2 m dew point temperature [K] | 0.342 (0.43%) | 0.052 (5.05%) | 0.030 | average [K] | 4 |
| | | | 0.100 (7.23%) | 0.055 | ITD latitude [°N] | 5 |
| 4 | MSLP [Pa] | 14.82 (0.47%) | 5.392 (9.62%) | 4.061 | SHL pressure [Pa] | 6 |
| | | | 0.145 (4.75%) | 0.095 | SHL boundary [°N] | 7 |
| 5 | column-integrated water vapor [kg m$^{-2}$] | 0.536 (0.33%) | 0.125 (7.04%) | 0.071 | average [kg m$^{-2}$] | 8 |
| 6 | high-level cloud cover [%] | 6.64 (8.06%) | 0.862 (1.22%) | 0.247 | average [%] | 9 |
| 7 | mid-level cloud cover [%] | 1.91 (1.53%) | 0.193 (1.71%) | 0.106 | average [%] | 10 |
| 8 | low-level cloud cover [%] | 1.60 (0.44%) | 0.223 (4.73%) | 0.128 | average [%] | 11 |
| 9 | u-wind at 600 hPa [m s$^{-1}$] | 0.436 (1.52%) | 0.069 (55.24%) | 0.088 | AEJ speed [m s$^{-1}$] | 12 |
| | | | 0.078 (5.81%) | 0.053 | AEJ latitude [°N] | 13 |
| 10 | u-wind at 200 hPa [m s$^{-1}$] | 0.454 (0.40%) | 0.116 (7.02%) | 0.059 | TEJ speed [m s$^{-1}$] | 14 |
| | | | 0.044 (17.62%) | 0.022 | TEJ latitude [°N] | 15 |
| 11 | v-wind at 600 hPa [m s$^{-1}$] | 0.177 (4.37%) | - | - | - | - |
| 12 | v-wind at 200 hPa [m s$^{-1}$] | 0.346 (1.34%) | - | - | - | - |

## 4.4 Investigation of parameter impacts

To ensure a structured progression from fundamental concepts to more complex analyses, the study first examines scalar QoIs, applying GSA and presenting the corresponding results. Subsequently, the influence of model parameters on QoIs and meteorological fields is investigated through parameter studies.

### 4.4.1 Global sensitivity analysis

To assess the magnitudes of model parameter impacts on scalar QoIs, GSA is applied as detailed in Section 2.5, specifically using the FAST method, with respect to the model parameters and their corresponding PDFs (Table 4.1). Since the surrogate models have been developed in the i. i. d. uniform input space, this input space can be directly utilized in GSA.

The results of GSA are shown in Fig. 4.2. For each QoI, the bar plots indicate the main and total effect sensitivity indices of the six uncertain model parameters. These results should be interpreted as relative contributions to the total variance of each QoI, meaning that comparing magnitudes between different QoIs is not meaningful. A comparison of absolute uncertainty contributions between different QoIs is inherently difficult or impossible, as they have different units. Overall, the main and total effect indices do not differ significantly, indicating that interactions between parameters are relatively weak. This justifies analyzing the influence of individual model parameters on QoIs separately, as done later in this study. However, interactions between parameters are expected to be larger for broader parameter ranges, where nonlinear effects may become more dominant.

Sensitivities of high- and mid-level cloud cover are generally dominated by the two deep-cloud parameters, the entrainment rate (*entrorg*) and the terminal fall velocity of ice (*zvz0i*). High-level clouds are strongly affected by entrainment, which can prevent convection from reaching high levels, whereas effects on mid-level clouds are minor. The fall velocity of ice controls the dissolution of high clouds but also has a dominant effect on mid-level clouds. Low-level clouds are affected by more parameters in a more complex way. As expected, below-cloud and boundary-layer parameters have a substantial effect at these altitudes. Particularly, the relative humidity threshold for onset of subcloud evaporation (*rhebc_land_trop*) and the surface area density of the evaporative soil surface (*c_soil*) exert the strongest influence on low clouds, while deep-cloud parameters play only a minor role (20 % combined).

Column-integrated water vapor is mostly influenced by deep-cloud parameters, similar to high clouds, although boundary-layer parameters also play a minor role. This suggests that column-integrated water vapor is more sensitive to interactions with mid- and high-level clouds than to evaporation and vertical mixing at low levels. Somewhat unexpectedly, 2 m temperature and 2 m dew point temperature are also predominantly influenced by deep-cloud parameters, with entrainment rate being the most significant factor. This suggests that entrainment processes induce substantial indirect effects beyond clouds. More intuitively, *c_soil* significantly affects surface-level thermodynamics.

Precipitation exhibits the most complex response, as it is influenced by all model parameters except for the convective area fraction used for computing evaporation below cloud base (*rcucov_trop*). While the impact of deep-cloud parameters is expected, the significant influence of boundary-layer parameters highlights the importance of low-level moisture for precipitation. The parameter *rhebc_land_trop* also exhibits a minor influence, as evaporation below cloud base directly affects surface rainfall.

The AEJ speed and position are most sensitive to *entrorg*, followed by *zvz0i*, suggesting that deep clouds play the dominant role, likely via their effects on baroclinicity and vertical momentum transport. Notably, the AEJ speed is the only variable with a significant contribution from *rcucov_trop*, possibly due to its location in the relatively dry Sahel, where evaporation below cloud base is substantial. The latitudes of the ITD and AEJ exhibit similar sensitivities, indicating a tight coupling between the two. The TEJ speed is primarily controlled by *zvz0i*, as it influences divergent outflow from convective anvils, which feeds the TEJ [74]. Interestingly, the TEJ position is also sensitive to entrainment and even boundary-layer parameters, showing the largest difference between total and main effects among all variables. The strength and latitude of the SHL are most sensitive to *entrorg*, which is surprising given the absence of deep clouds over most of the Sahara. A possible explanation is that entrainment influences free-tropospheric water vapor content, which strongly affects longwave cooling in dry regions [108]. Finally, the latitude of the precipitation maximum is primarily controlled by *rhebc_land_trop* (~65 % contribution) with minor contributions from all other parameters. This behavior is in stark contrast to precipitation amount and almost all other QoIs in Fig. 4.2. Given the strong gradient in absolute and relative humidity across the Sahel, this finding suggests that modifying the onset of subcloud evaporation in the model is a powerful mechanism for shifting the entire rain belt meridionally. This result may help explain variability in rain belt position observed in model intercomparison projects [30].

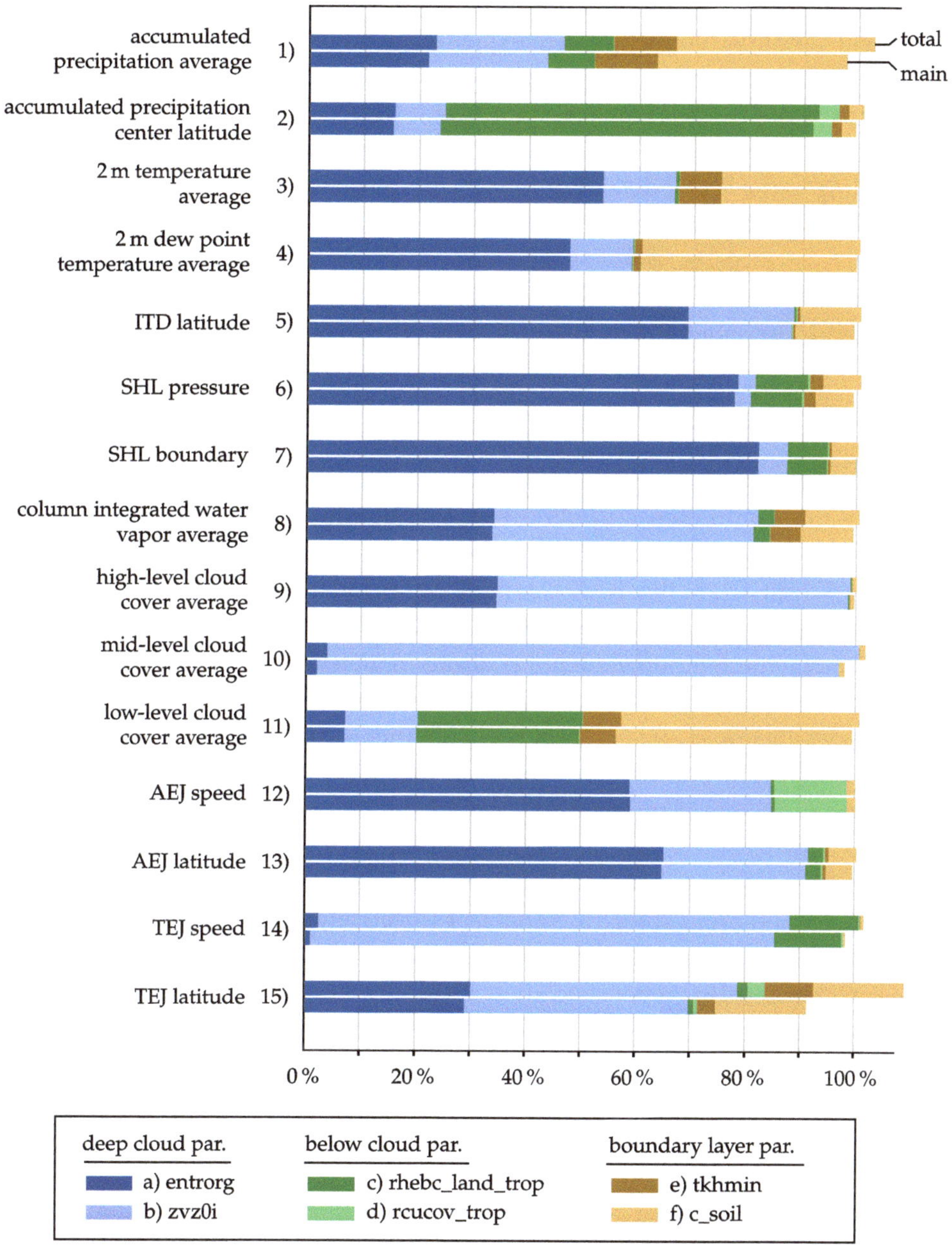


**Figure 4.2:** Total and main effect sensitivity indices of model parameters for all QoIs.

### 4.4.2 Parameter studies

The surrogate models for QoIs and meteorological fields are used to analyze the effects of parameter variations on the respective quantities. Since surrogate model predictions depend on all six parameters simultaneously, a full graphical representation of these relationships is not feasible. Instead, a one-at-a-time variation approach is applied, where a single parameter is varied while all others are held at their mean values, corresponding to the ICON default settings. This approach is justified by the relatively low parameter interactions identified in Section 4.4.1.

Figure 4.3 illustrates the relationships between each model parameter and each QoI. Given the weak parameter interactions, choosing different fixed values instead of mean values would primarily result in vertical shifts of the depicted curves. The predictive variance from GPR (Eq. 2.9) is represented by shaded areas around the curves.

Figure 4.4 presents reference fields for all meteorological variables, obtained by averaging ICON simulation outputs over all 60 training points and the entire evaluation period (August 2016–2019). These reference fields are preferred over the simulation output from the ICON default parameter settings, as they provide a more robust representation of the system's average state. A continuous graphical visualization of parameter-dependent variations in meteorological fields would require three-dimensional plots, which are impractical. However, given the predominantly monotonic dependencies observed in the QoIs (Fig. 4.3), it is reasonable to consider the average impact of parameter changes on meteorological fields. To quantify this effect, the mean change in meteorological fields is computed for parameter variations from $-\sigma$ to $\sigma$, where $\sigma$ is the standard deviation of the corresponding PDF. This is done by subtracting the surrogate field for parameter $-\sigma$ from that for parameter $\sigma$ for each parameter and meteorological variable, respectively. The resulting fields, referred to as spatial variability fields, indicate the regions where a meteorological variable increases or decreases in response to an increase in a specific model parameter. Spatial variability fields for all three parameter groups are shown in Figs. 4.5, 4.6, and 4.7, respectively. Since the full dependence structure cannot be visualized in such plots, an interactive tool has been developed as part of this research. This tool, accessible at mattfis.github.io/wam-simulations [159], is based on the results from PCR and enables modelers to explore the spatial effects of combined parameter changes. Figure 4.8 shows a screenshot from this website for a specific parameter configuration.

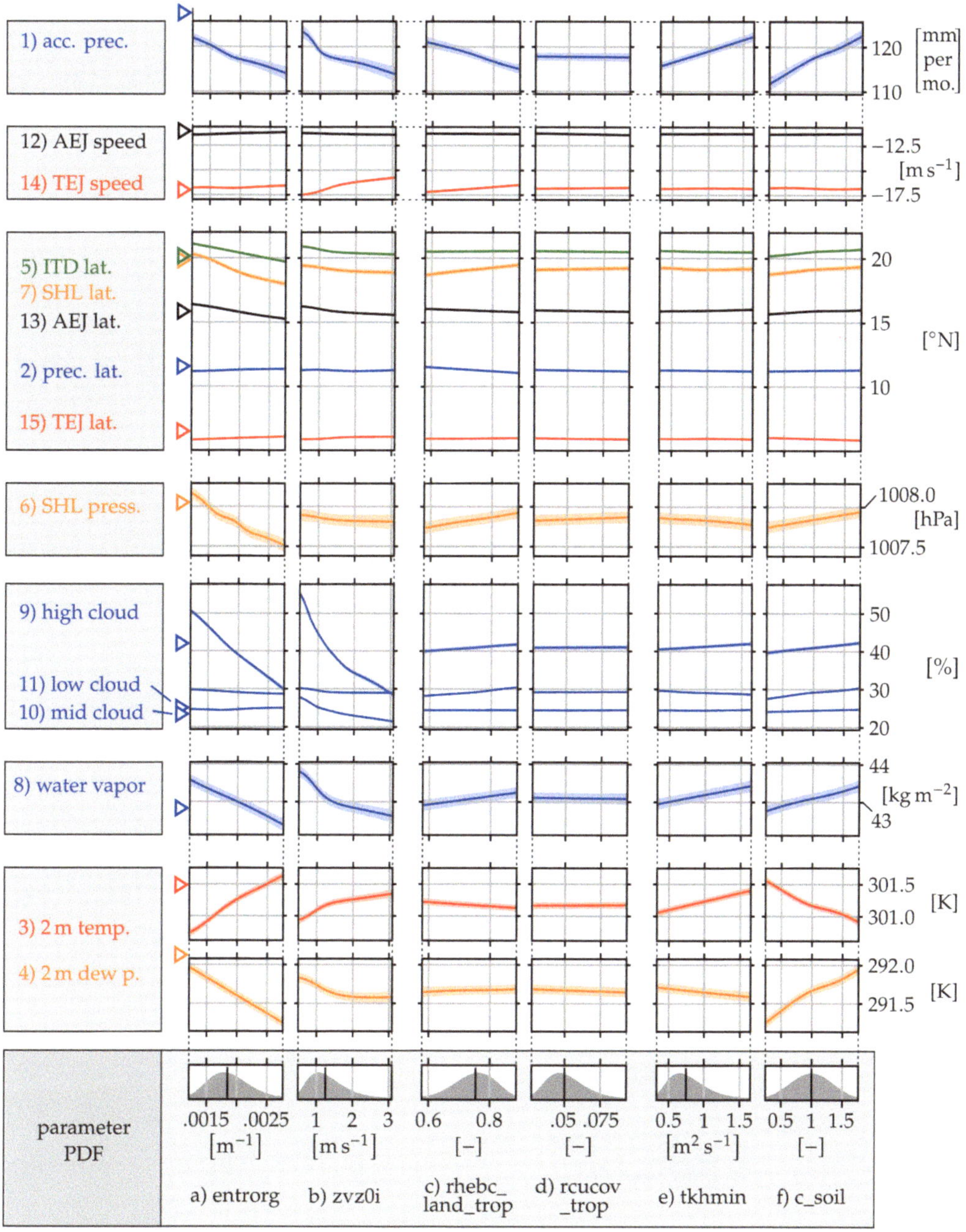


**Figure 4.3:** Dependencies of all QoIs (ordinate) with respect to the six selected uncertain model parameters (abscissa), respectively. The shaded area around curves illustrates the predictive variance (see Eq. 2.9). In each plot, only one model parameter is varied while all others are fixed at their default value. The PDFs of the model parameters, including their default values, are shown at the bottom. Triangles on the left indicate the values computed from the reference data (Section 4.2.1).

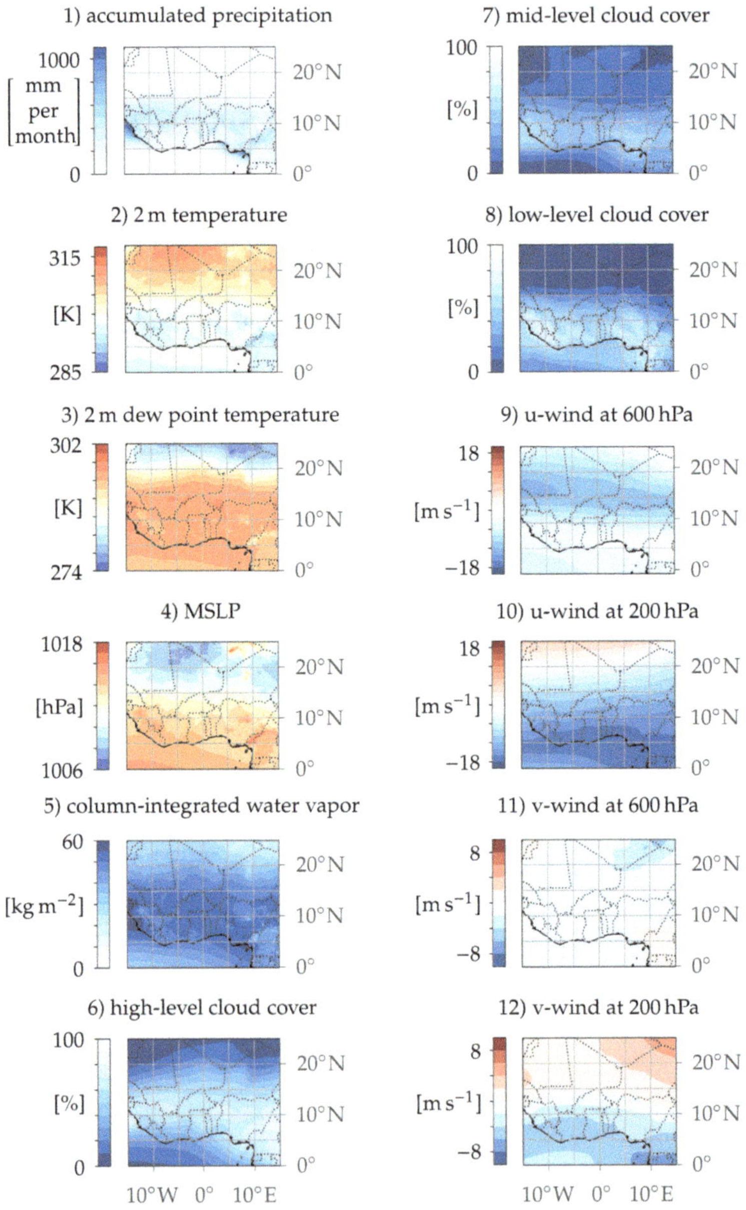


**Figure 4.4:** Average of selected meteorological fields over the evaluation period (Augusts 2016–2019), computed from the ICON simulation output and averaged over all 60 training points. All color scales are linear.

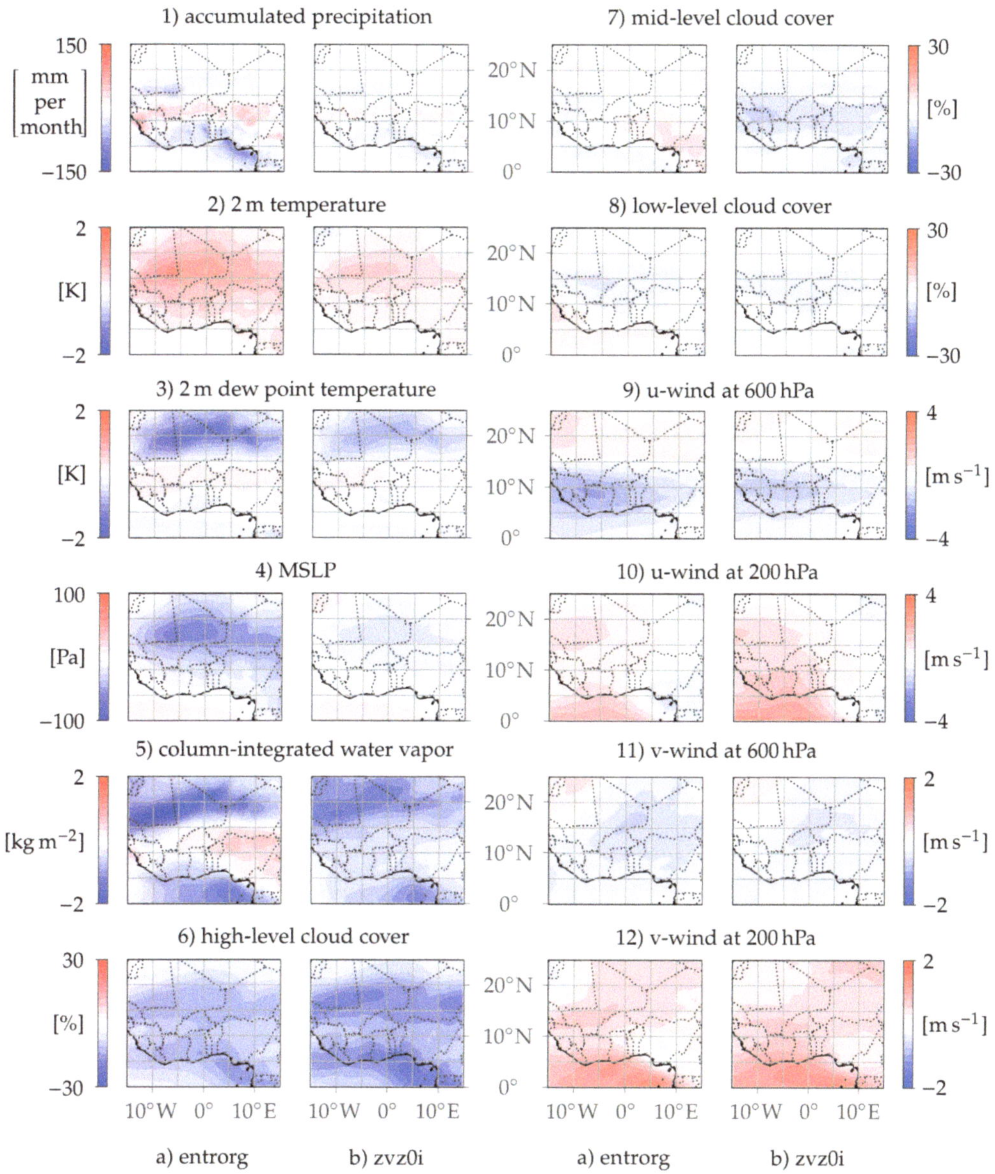


**Figure 4.5:** Spatial variability fields for *entrorg* and *zvz0i*. Colors indicate an increase (red) or decrease (blue) in the meteorological variable in response to an increase in the respective model parameter from $-\sigma$ to $\sigma$ of its PDF. All color scales are linear.

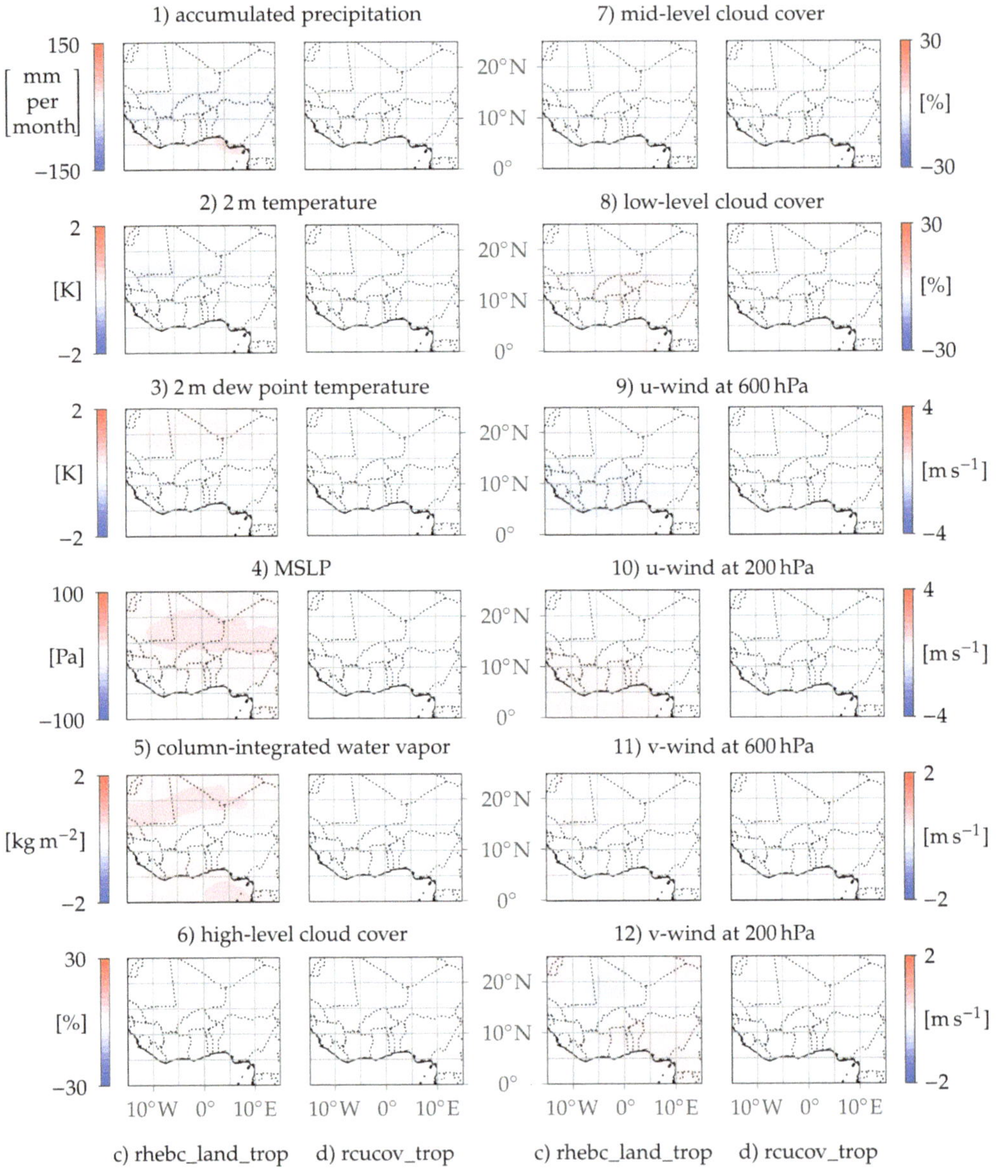


**Figure 4.6:** Spatial variability fields for *rhebc_land_trop* and *rcucov_trop*. Colors indicate an increase (red) or decrease (blue) in the meteorological variable in response to an increase in the respective model parameter from $-\sigma$ to $\sigma$ of its PDF. All color scales are linear.

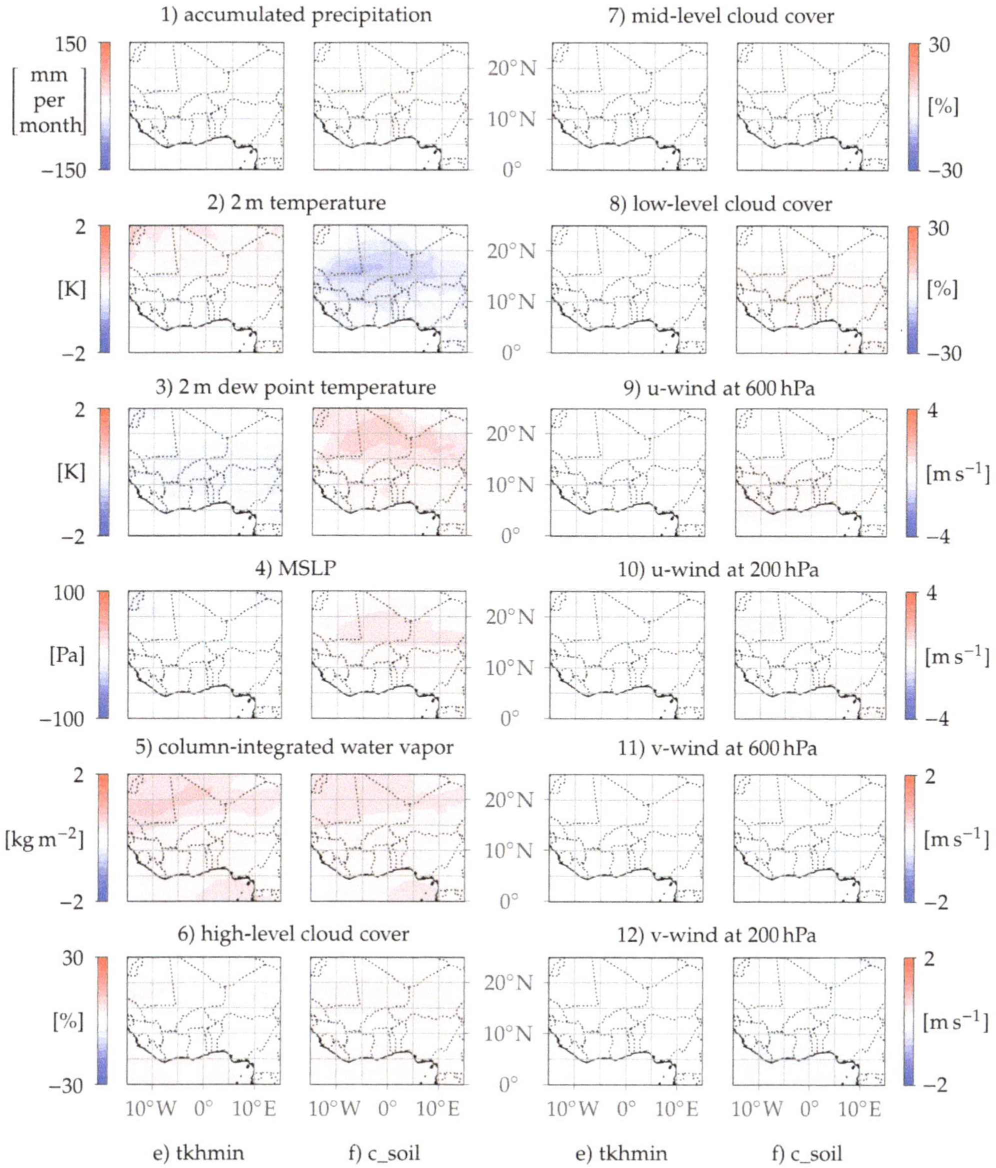


**Figure 4.7:** Spatial variability fields for *tkhmin* and *c_soil*. Colors indicate an increase (red) or decrease (blue) in the meteorological variable in response to an increase in the respective model parameter from $-\sigma$ to $\sigma$ of its PDF. All color scales are linear.

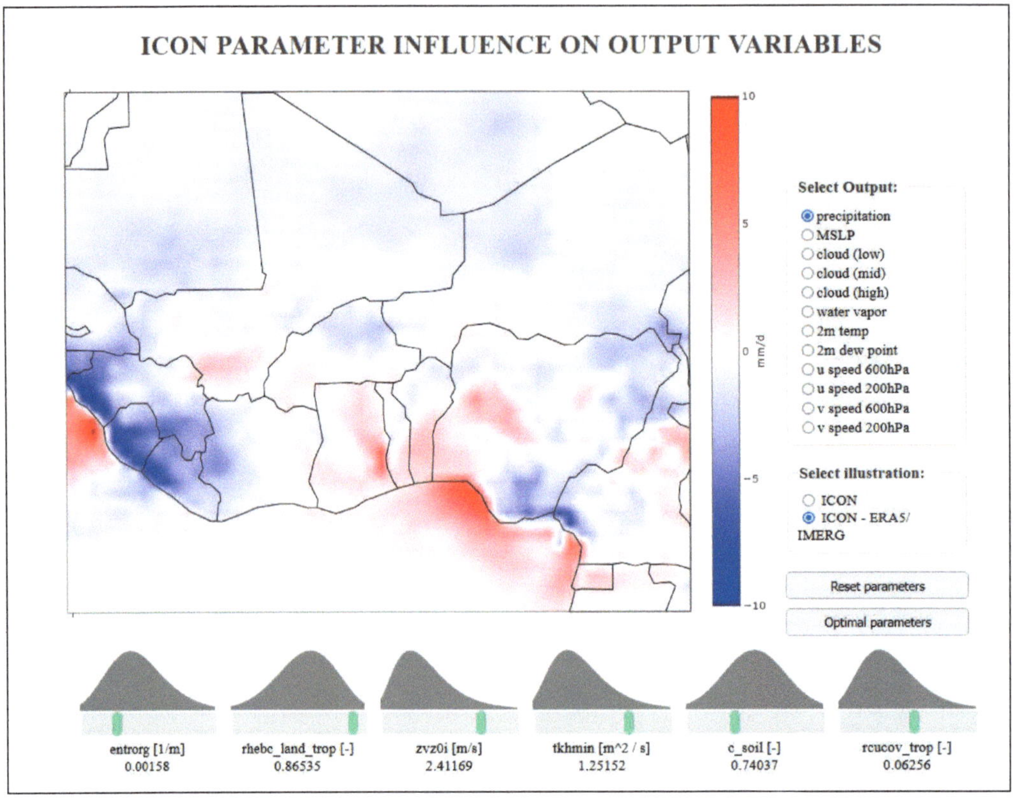


**Figure 4.8:** Screenshot of the developed online tool [159], accessible at mattfis.github.io/wam-simulations

## Deep-cloud parameters

The effect of the investigated deep-cloud parameters, entrainment rate (*entrorg*) and terminal fall velocity of ice (*zvz0i*), is considerably greater for most QoIs than that of other parameters, as evident from Fig. 4.2 (blue-colored bars) and Fig. 4.3. Since both parameters directly affect cloudy regions, signals outside the rain belt arise primarily due to indirect effects.

As shown in Fig. 4.3, an increase in *entrorg* leads to a decrease in 2 m dew point, column-integrated water vapor, high-level cloud cover, and precipitation, suggesting an overall drying of the WAM system. This is accompanied by an increase in 2 m temperature and lower pressure in the SHL. Additionally, a consistent southward shift of the northern WAM features (ITD, SHL boundary, and AEJ) is observed, while the southern features (precipitation center and TEJ) remain at their latitudes. The strengths of the jets as well as low- and mid-level cloud cover are barely affected.

Figure 4.5a (i. e. the 1st and 3rd column from the left) shows the corresponding results on a horizontal map. Increasing *entrorg* reduces precipitation north and south of the rain belt, as expected from Fig. 4.3, but surprisingly slightly increases precipitation within a narrow strip through the rain belt (Fig. 4.5a1). This increase may be attributed to rainfall becoming concentrated in regions where ambient conditions are most favorable, while higher entrainment suppresses precipitation in more marginal areas. It is also possible that the southward shift of the AEJ (Fig. 4.3) alters the distribution of low-level wind shear, which plays a crucial role in convective organization [22]. Despite the local precipitation increase, high clouds decrease across the entire domain by up to 25 % (Fig. 4.5a6), though less so over the rain belt, where they climatologically maximize (Fig. 4.4.6). This suggests that weaker convective systems are suppressed, and rainfall is generated more efficiently by fewer, more intense systems. A higher *entrorg* also results in an increase in mid-level clouds in the southeastern parts of the domain (Fig. 4.5a7), while oceanic and western land areas show a slight decrease. Given that entrainment reduces convective instability, it is plausible that clouds are retained at mid-levels in marginally unstable regions, though the exact mechanisms behind the spatial distribution remain unclear. Regarding low-level cloud cover (Fig. 4.5a8), increased entrainment leads to a widespread reduction over the Sahel, suggesting that the northern edge of the low-cloud zone over southern West Africa (see Fig. 4.4.8) retreats southward, while cloud cover over the ocean and coastal areas increases. This shift may be linked to the overall southward displacement of several WAM features, as shown in Fig. 4.3. The high sensitivity of high-level clouds largely determines the signal in total cloud cover (not shown).

The column-integrated water vapor (Fig. 4.5a5) decreases in and around regions with reduced precipitation and increases (or remains unchanged) in wetter regions, particularly in the southeast, where an increase in mid-level clouds is also observed (Fig. 4.5a7). Over the Sahara, drying is also pronounced at the surface (2 m dew point, Fig. 4.5a3) but less so farther south. The decrease may be a result of reduced rainfall and evaporation, coupled with a southward-shifted monsoon circulation. Conversely, the slight increase in water vapor over the rain belt is likely a direct consequence of enhanced rainfall. The overall reduction in cloud cover, precipitation, and evaporation leads to surface warming across nearly the entire land area of the domain (Fig. 4.5a2), which is associated with a lower MSLP due to thermal expansion (Fig. 4.5a4). The maximum MSLP reduction occurs south of the climatological SHL center (Fig. 4.4.4), contributing to a southward shift. Additionally, altered temperature advection linked to the southward shift of the ITD (see Fig. 4.3) may also play a role.

As already noted in the discussion of Fig. 4.3, the sensitivity of the zonal jets to *entrorg* is relatively weak. The most systematic response is a clear southward shift in zonal wind at 600 hPa (Fig. 4.5a9), with a decrease of several $m\,s^{-1}$ south of the climatological axis (Fig. 4.4.9). In the meridional direction (Fig. 4.5a11), an overall strengthening of the climatological northerlies is observed (Fig. 4.4.11), suggesting a stronger shallow monsoon circulation, consistent with the more pronounced SHL (Fig. 4.5a4). At the TEJ level (200 hPa), the broad climatological easterlies (Fig. 4.4.10) are slightly weakened by higher entrainment, except in the southeastern corner of the domain (Fig. 4.5a10). In the meridional direction (Fig. 4.5a12), the reduction in rainfall over the Guinea Coast is associated with a weakening of the northerly divergent outflow toward the equatorial Atlantic (Fig. 4.4.12), which likely contributes to a weaker TEJ in the west [74]. At the same time, the outflow into the Northern Hemisphere is slightly enhanced, shifting the relative importance of the two deep monsoonal overturning cells. Given the substantial decrease in high-level cloud cover (Fig. 4.5a6), it is also plausible that radiative cooling in the upper troposphere increases [126], which would contribute to a weaker monsoon cell, consistent with a Gill-type circulation response to reduced off-equatorial heating [38].

Comparing the effect of increased entrainment with that of a higher terminal fall velocity of ice, several similarities emerge, despite the fundamentally different microphysical processes involved. Regarding the overall effects presented in Fig. 4.3 (first two columns), most signals are consistent in sign and, in many cases, in magnitude. The most notable differences include a northward shift of the TEJ with increasing *zvz0i*, a weaker impact on SHL strength, a decrease in mid-level clouds, and a smaller effect on the 2 m dew point. Examining the corresponding horizontal distributions (2nd and 4th columns in Fig. 4.5), a striking similarity in spatial patterns is evident, albeit with some differences in magnitude. For instance, the signal in high-level cloud cover (Fig. 4.5b6) is stronger, as it is directly influenced by ice particles, while the effects on surface temperature, dew point, MSLP, and low-level cloud cover (Figs. 4.5b2, b3, b4, b8) are weaker, as these are predominantly indirect responses. The most striking difference is the absence of an anomalous response in the southeastern part of the domain. Here, the effects of higher *zvz0i* align more closely with those observed elsewhere, i. e. implying fewer mid-level clouds (due to faster dissipation), reduced column-integrated water vapor, and a weaker or unchanged TEJ (Figs. 4.5b5, b7, b10). Other circulation-related changes are almost identical (compare Figs. 4.5a9, a11, a12 with Figs. 4.5b9, b11, b12). The impact of a higher *zvz0i* on precipitation closely resembles that of *entrorg*, but with a lower amplitude.

**Below-cloud parameters**

The investigated below-cloud parameters, namely the relative humidity threshold for the onset of evaporation (*rhebc_land_trop*) and the convective area fraction used for computing evaporation (*rcucov_trop*), affect cloudy areas only. Consequently, their effects outside the rain belt are largely indirect. Their impact on almost all QoIs (Figs. 4.2, green-colored bars, and Fig. 4.3) and meteorological fields (Fig. 4.6) is considerably smaller compared to the deep-cloud parameters discussed in the previous subsection. Increased evaporation leads to cooler subcloud layers, which results in greater negative buoyancy and enhanced lateral acceleration relative to adjacent grid cells, resembling intensified cold pools. However, given the 13 km grid spacing used in the simulations, this process – particularly the triggering of new storms by cold pools – may not be fully resolved. Therefore, the findings provide only limited insights into the actual role of cold pools in the monsoon system and the potential benefits of a cold pool parameterization.

The most prominent signals in Fig. 4.2 are those for low-level cloud cover and precipitation latitude (both *rhebc_land_trop*), followed by precipitation amount, TEJ speed, SHL latitude and intensity (all *rhebc_land_trop*) and, to a lesser extent, AEJ speed (*rcucov_trop*). Examining the dependencies of the QoIs in Fig. 4.3 reveals that allowing evaporation at higher relative humidity in the model (i. e. increasing *rhebc_land_trop*) suppresses precipitation and induces a slight southward shift of the rain belt. This shift is driven by a precipitation decrease over the Sahel (Fig. 4.6c1), where cloud bases are higher and subcloud relative humidity is climatologically near the threshold. At the same time, MSLP increases across the northern and central parts of the domain (Fig. 4.6c4), which is associated with a weakening and slight northward shift of the SHL (Fig. 4.3). The increase in subcloud evaporation also corresponds to slightly more low-level clouds over inland areas south of the Sahara (Fig. 4.6c8), whereas 2 m temperature and dew point remain largely unchanged (Fig. 4.6c2 and c3). The small temperature response near the surface could result from reduced surface evaporation due to lower rainfall and soil moisture, which might compensate for the reduced radiative heating from increased low-level clouds and the enhanced subcloud evaporative cooling. Additionally, changes in temperature advection associated with the weaker SHL may also contribute. Interestingly, increasing *rhebc_land_trop* also influences column-integrated water vapor (Fig. 4.6c5), but predominantly outside the rain belt, particularly over the Gulf of Guinea and Mauritania. The increase in moisture in these areas aligns with weaker overturning circulations due to suppressed precipitation, possibly leading to a redistribution of atmospheric moisture. Some indications of this can also be seen in the 200 hPa wind field, which exhibits a slight reduction in northerly outflow over Nigeria (Fig. 4.6c12) and a weakening of the TEJ (Fig. 4.6c5). Other fields for *rhebc_land_trop* and all fields for *rcucov_trop* (Fig. 4.6d) show no significant signals, consistent with Figs. 4.2 and 4.3.

**Boundary-layer parameters**

The effects of the scaling factor for minimum vertical diffusion of heat and moisture (*tkhmin*) and the surface area density of the evaporative soil surface (*c_soil*) are less pronounced compared to those of the deep-cloud parameters. The strongest sensitivities are observed for near-surface QoIs, including low-level cloud cover, 2 m temperature, and 2 m dew point, as well as for integrated quantities such as column-integrated water vapor and precipitation (Fig. 4.2, brown-colored bars).

For higher values of *tkhmin*, moisture is transported more efficiently upward, resulting in an increase in column-integrated water vapor across most of the domain (Fig. 4.7e5), as also evident in Fig. 4.3. This enhanced vertical moisture transport leads to an increase in high clouds in various regions (Fig. 4.7e6), while mid- and low-level cloud cover remain largely unchanged. The stronger vertical transport of moisture from the surface also enhances evaporation, which is reflected in lower 2 m dew points over the Sahel (Fig. 4.7e10). The impact on 2 m temperature (Fig. 4.7e2) is minimal, except for an increase over the Sahara, where longwave warming due to higher column moisture and/or the mixing of warm air from above the boundary layer inversion may contribute. The enhanced vertical exchange of moisture leads to a slight increase in accumulated precipitation (Fig. 4.3), though this effect is not clearly visible in the spatial field (Fig. 4.7e1). Similarly, MSLP exhibits a slight strengthening of the SHL (Fig. 4.3), yet the spatial field shows little signal (Fig. 4.7e4). Other fields are only marginally affected or remain unchanged by *tkhmin*.

In contrast, *c_soil* directly enhances surface evaporation, resulting in a significant increase in 2 m dew point (Fig. 4.7f3) and a decrease in 2 m temperature (Fig. 4.7f2) over nearly all land areas. This effect is also clearly evident in the overall dependencies shown in Fig. 4.3. Column-integrated water vapor (Fig. 4.7f5) also increases, particularly north and south of the rain belt, where enhanced precipitation (Fig. 4.7f1, see also Fig. 4.3) likely removes some of the additional moisture. This increase in precipitation is also accompanied by a slight enhancement of high-level clouds (Fig. 4.7f6). As increased surface latent heat fluxes over land areas moisten the boundary layer, low-level cloud cover rises over the southern parts of West Africa (Fig. 4.7f8), which may further intensify near-surface cooling. This cooling, in turn, leads to an increase in MSLP (Fig. 4.7f4), contributing to a weakening of the SHL and a northward shift of both the

SHL and ITD (Fig. 4.3). Since an increase in *c_soil* promotes stronger convection and enhanced precipitation (Fig. 4.3), i. e. an overall strengthening of the WAM system, the 200 hPa outflow from the rain belt to the south is intensified (Fig. 4.7f12), leading to an acceleration of the TEJ (Fig. 4.7f10). In contrast, the AEJ remains only weakly affected.

**Summary of parameter impacts**

Although the magnitude of the impact of individual model parameters varies considerably, most parameters exhibit distinct effects on multiple aspects of the system. These effects are schematically illustrated for the four most influential parameters in Fig. 4.9. The key findings can be summarized as follows:

- The entrainment rate (*entrorg*) and terminal fall velocity of ice (*zvz0i*) exert the strongest influence on the WAM system (see Fig. 4.9a). Higher values of these parameters reduce cloud cover and precipitation, particularly north and south of the rain belt across West Africa. However, for *entrorg*, precipitation increases along a narrow strip within the rain belt, likely benefiting from suppressed rainfall in adjacent areas. Larger values of both parameters contribute to a stronger SHL, characterized by warmer and drier conditions over the Sahara, as well as a stronger shallow overturning circulation. This is accompanied by a southward shift of the ITD and AEJ, while the TEJ weakens.

- The parameters *rhebc_land_trop* and *rcucov_trop* regulate evaporation below the cloud base in the tropics and have a weaker overall impact on the WAM. Increasing *rhebc_land_trop* (Fig. 4.9b) leads to reduced precipitation and increased low-level cloud cover. This appears to weaken monsoon overturning, as reflected by a weaker SHL and moister atmospheric columns in subsidence regions over the northwestern Sahara and the Gulf of Guinea, though its effects on the AEJ and TEJ are minimal. Changing *rcucov_trop* has only marginal effects.

- The scaling factor for vertical diffusion of heat and moisture (*tkhmin*) influences moisture exchange between the boundary layer and the free troposphere. Increasing this parameter leads to higher column-integrated water vapor and more high- and mid-level clouds, though precipitation remains largely unchanged. The evaporative soil surface (*c_soil*) also increases column-integrated water vapor and cloud cover, but primarily enhances low-level clouds, resulting in a small precipitation increase along the southern side of the rain belt (see Fig. 4.9c). Near-surface temperatures decrease due to enhanced evaporation, while 2 m dew point temperature and MSLP increase, shifting the SHL northward. The impacts of both *tkhmin* and *c_soil* on the AEJ and TEJ are minor.

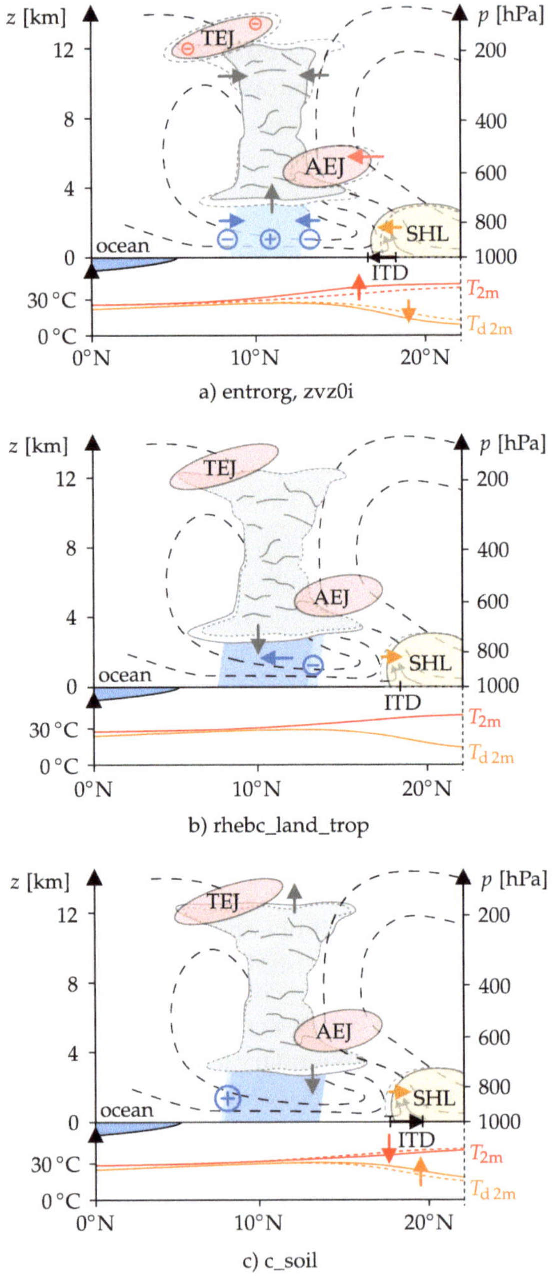


**Figure 4.9:** Illustration of the qualitative effects on the WAM system due to an <u>increase</u> in respective model parameters with the strongest impacts.

## 4.5 Multi-objective optimization for parameter identification

In this section, the methods introduced in Section 3.2 are applied to determine optimal model parameters, optimizing either meteorological fields or scalar QoIs using the surrogate models obtained in Section 4.3. These methods provide insight into the sensitivity of optimization results to variations in weights assigned to individual objectives when formulating a reduced single-objective optimization problem from multiple objectives. To better interpret the optimization results and place them in a broader context, it is first necessary to identify the meteorological changes being targeted. For this purpose, a detailed comparison is conducted between ICON simulation results and reference data, against which the ICON model is optimized.

### 4.5.1 Comparison of ICON output to the reference data

Notable discrepancies between the ICON model output and the reference data (GPM IMERG and ERA5) are examined to identify constraints and opportunities for the optimization process. In Fig. 4.3, alongside the dependencies of QoIs describing the ICON model output, the triangles on the left indicate the reference values that serve as optimization targets. Additionally, Fig. 4.10 presents the differences between the ICON model output, using default model parameters, and the reference data fields (see Section 4.2.1) for each considered meteorological field. The following discussion focuses on the most critical variables, determined by the assigned weights (Table 4.6).

With respect to precipitation, the domain-mean rainfall in ICON simulations is approximately 10 % lower than observed in GPM IMERG data (Fig. 4.3, first row). ICON underestimates rainfall in coastal regions (southwestern West Africa, Niger Delta) and mountainous areas (Guinea Highlands, Cameroon Line) as well as over the Sahara, while exhibiting a slight overestimation in central West Africa (Fig. 4.10.1). This overestimation results in a small northward shift of the precipitation center (Fig. 4.3, seventh row). The underestimation in coastal and mountainous regions suggests that the model struggles to capture rain enhancement due to topographic features. The northward shift of the precipitation center may be linked to the use of convection parameterization in the ICON simulations. Kniffka, Knippertz, and Fink [63] demonstrated that models employing parameterized convection tend to misrepresent the northward migration of the rain belt, leading to reduced rainfall in the Sahel. A follow-up study by Kniffka et al. [64] found that total rainfall from ICON simulations with parameterized convection is largely consistent with station data in southern West Africa, whereas GPM IMERG data show a negative bias relative to the stations, highlighting uncertainties in the observational reference. Overall, these findings leave open the question of whether adjusting model parameters can resolve the rainfall deficiencies or whether improvements in topographic representation are the primary requirement.

The dominant pattern in the MSLP differences between ICON and ERA5 is a zonal dipole, with slightly higher pressure in the east and lower pressure in the northwest of the domain (Fig. 4.10.4). This results in a deeper SHL in ICON (Fig. 4.3). The pressure pattern is strongly correlated with 2 m temperature, where higher pressure values correspond to lower surface temperatures (Fig. 4.10.2). At the same time, ICON exhibits lower 2 m dew point values, except for a region near the Algeria-Mali-Niger border (Fig. 4.10.3), though this has little effect on the mean ITD latitude (Fig. 4.3). Despite this, ICON simulates higher column-integrated water vapor across most of West Africa (Fig. 4.10.5) and on average (Fig. 4.3), suggesting differences in the vertical distribution of water vapor between ICON and ERA5.

This is partially reflected in cloud cover differences at low, mid, and high levels. Regarding low-level cloud cover, ICON simulations exhibit a pronounced positive bias across nearly the entire rain belt relative to ERA5 reanalysis, with a minor negative bias over the equatorial Atlantic Ocean (Fig. 4.10.8). This suggests a northward extension of the low-level cloud belt in ICON. The domain-averaged low-level cloud cover in ICON reaches 30 %, significantly exceeding the 25 % in ERA5 data (Fig. 4.3). Such a large discrepancy is somewhat unexpected, as it is primarily accompanied by lower 2 m dew point temperatures, indicating lower absolute humidity. This suggests that calibrating the model using the selected uncertain parameters will not be straightforward, as achieving agreement with observations would likely require complex interactions within the WAM system.

Accurately representing low-level cloud cover has been identified as a major challenge in previous studies, with significant discrepancies observed between various models and observational data [45, 64]. Kniffka et al. [64] demonstrated that low-level cloud cover in ICON with parameterized convection deviates by only 2 % from station data, raising questions about the reliability of ERA5 cloud estimates. Mid-level clouds are mostly enhanced in ICON over the rain belt area but reduced elsewhere (Fig. 4.10.7). In contrast, high-level clouds are generally less widespread in ICON, except for specific regions in Mali and Niger (Fig. 4.10.6). This partially contradicts the findings of Kniffka, Knippertz, and Fink [63], who reported an overestimation of high-level clouds in ICON.

Moderate differences are also observed in the circulation patterns. The AEJ is

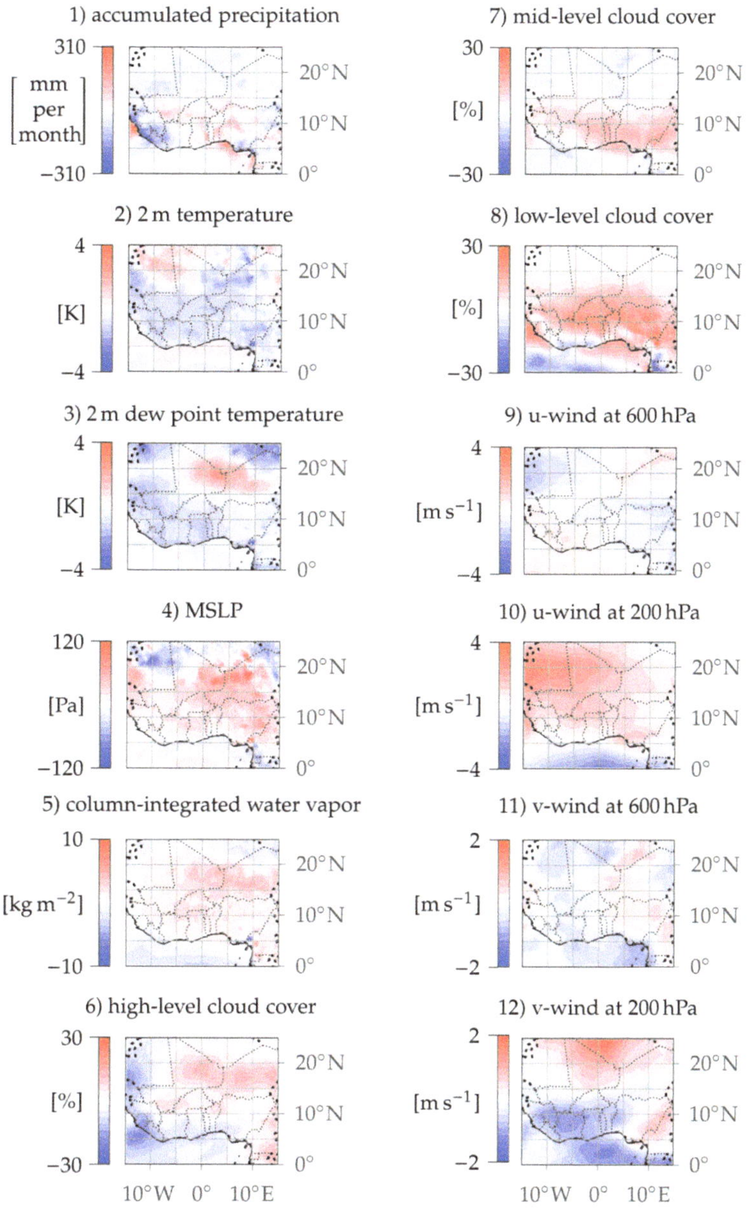


**Figure 4.10:** Difference (ICON − reference) between ICON (default parameters) and reference data (ERA5 reanalysis, and GPM IMERG for precipitation) for averaged meteorological fields.

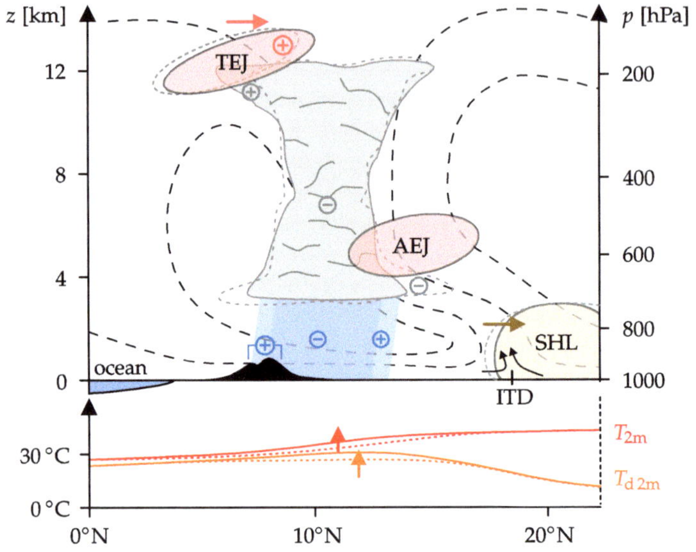


**Figure 4.11:** Desired adjustments of WAM characteristics based on the default ICON model output to better approximate reference data. See Fig. 4.10c for quantitative maps.

slightly stronger in ICON, whereas the TEJ is weaker and shifted southward (Fig. 4.3). The horizontal distribution of differences between ERA5 and ICON reveals that the subtropical jet is also displaced southward in ICON (Fig. 4.10.10), while differences in zonal wind at 600 hPa remain relatively small (Fig. 4.10.9). Finally, meridional wind fields exhibit only moderate differences, characterized by a fairly patchy pattern at both 600 hPa and 200 hPa (Figs. 4.10.11 and 12).

For a clearer overview, Fig. 4.11 summarizes the necessary adjustments in WAM characteristics in ICON simulations to improve agreement with reference data. Precipitation should increase overall, particularly in mountainous regions near the Atlantic coast. The SHL should weaken and shift northward. Cloud cover should decrease at low and mid levels, while high-level cloud cover should increase south of the rain belt. 2 m temperature and 2 m dew point temperature should increase on average over southern West Africa. The TEJ should be slightly strengthened, while the AEJ should remain unchanged.

### 4.5.2 Optimization objectives

As a first step, separate objective functions for the QoIs and meteorological fields are defined with respect to the model parameters **u** in the i. i. d. uniform input space. These objective functions are formulated such that their minimization leads to an improvement in the considered meteorological variables relative to the reference data. For the QoIs, the goal is to adjust the model parameters so that the QoI values describing the ICON model output closely match the QoI values from the reference data, as represented by the triangles in Fig. 4.3. For the meteorological fields, the objective is to modify the parameters such that the differences illustrated in Fig. 4.10 are minimized, ideally canceling out the deviations as much as possible, leading to a white image.

The individual objective functions for the QoIs $f_{\mathrm{QoI},j}(\mathbf{u})$ and meteorological fields $f_{\mathrm{field},j}(\mathbf{u})$ are formulated as the square errors between the surrogate model and the reference data

$$f_{\mathrm{QoI},j}(\mathbf{u}) = (\hat{y}_{\mathrm{QoI},j}(\mathbf{u}) - y_{\mathrm{ref},j})^2 \quad (j = 1 \ldots 15)\,,$$
$$f_{\mathrm{field},j}(\mathbf{u}) = \sum_{k=1}^{50}\sum_{l=1}^{60}\left(\hat{M}_j^{kl}(\mathbf{u}) - M_{\mathrm{ref},j}^{kl}\right)^2 \quad (j = 1 \ldots 12)\,,$$

where, for the meteorological fields, the average over all 50×60 grid points is considered.

The aim of this study is to solve a MOO problem involving 15 QoIs or 12 meteorological field objectives. Given the complexity of the high-dimensional objective space, identifying Pareto fronts is considered impractical due to the substantial computational resources required and the challenges in interpreting the results. To address this, the MOO problem is simplified by reducing it to multiple single-objective optimization problems, a common approach in meteorological literature [10]. The reduced objectives are formulated by combining components of the original MOO problem. By applying a weighted sum approach, a relative importance is assigned to each individual objective, ensuring a structured and computationally feasible optimization process.

The total reduced objective functions using the QoIs and the meteorological fields are defined as

$$f_{\mathrm{QoI}}(\mathbf{u}) = \sum_{j=1}^{15} \frac{w_{\mathrm{QoI},j}}{\sigma_{\mathrm{QoI},j}} f_{\mathrm{QoI},j}(\mathbf{u})\,,$$
$$f_{\mathrm{field}}(\mathbf{u}) = \sum_{j=1}^{12} \frac{w_{\mathrm{field},j}}{\sigma_{\mathrm{field},j}} f_{\mathrm{field},j}(\mathbf{u})\,,$$

where $w_{\mathrm{QoI},j}$ and $w_{\mathrm{output},j}$ are the weights for the individual objectives that have to be specified in advance.

**Table 4.6:** Default weight configuration $\mathbf{w}_0$ of the objectives in MOO, including the spread for the *weight uncertainty* method in MOO.

| j | meteorological field [unit] | weight range | | QoI [unit] | j |
|---|---|---|---|---|---|
| 1 | accumulated precipitation [mm mo$^{-1}$] | 10 ± 50 % | 10 ± 50 % | average[1] [mm mo$^{-1}$] | 1 |
| | | | 0.1 ± 50 % | center latitude [°N] | 2 |
| 2 | 2 m temperature [K] | 5 ± 50 % | 5 ± 50 % | average[1] [K] | 3 |
| 3 | 2 m dew point temperature [K] | 5 ± 50 % | 5 ± 50 % | average[1] [K] | 4 |
| | | | 0.1 ± 50 % | ITD latitude [°N] | 5 |
| 4 | MSLP [Pa] | 5 ± 50 % | 5 ± 50 % | SHL pressure [Pa] | 6 |
| | | | 0.1 ± 50 % | SHL boundary [°N] | 7 |
| 5 | column-integrated water vapor [kg m$^{-2}$] | 5 ± 50 % | 5 ± 50 % | average[1] [kg m$^{-2}$] | 8 |
| 6 | high-level cloud cover [%] | 1 ± 50 % | 1 ± 50 % | average[1] [%] | 9 |
| 7 | mid-level cloud cover [%] | 1 ± 50 % | 1 ± 50 % | average[1] [%] | 10 |
| 8 | low-level cloud cover [%] | 3 ± 50 % | 3 ± 50 % | average[1] [%] | 11 |
| 9 | u-wind at 600 hPa [m s$^{-1}$] | 0.1 ± 50 % | 0.1 ± 50 % | AEJ speed [m s$^{-1}$] | 12 |
| | | | 0.1 ± 50 % | AEJ latitude [°N] | 13 |
| 10 | u-wind at 200 hPa [m s$^{-1}$] | 0.1 ± 50 % | 0.1 ± 50 % | TEJ speed [m s$^{-1}$] | 14 |
| | | | 0.1 ± 50 % | TEJ latitude [°N] | 15 |
| 11 | v-wind at 600 hPa [m s$^{-1}$] | 0.1 ± 50 % | - | - | - |
| 12 | v-wind at 200 hPa [m s$^{-1}$] | 0.1 ± 50 % | - | - | - |

[1] average value over all grid points of the study domain (15°W to 15°E, 0°N to 25°N)

The considered QoIs and meteorological fields, along with the default weight combination $\mathbf{w}_0$, are presented in Table 4.6. Precipitation is assigned the highest importance, reflecting its pivotal role in the WAM system and its relevance for user applications. Temperature, dew point temperature, pressure, and column-integrated water vapor are assigned secondary importance, as they are essential meteorological variables but remain subordinate to precipitation. Cloud cover is given a lower weight, with greater emphasis on low-level clouds. Finally, wind speeds receive the smallest weights but are still considered, ensuring that all quantities contribute to the optimization, with some acting more strongly while others serve as a form of regularization.

To ensure that variable weighting is meaningful and significant, the meteorological variables are normalized based on their variability. For this purpose, the standard deviations $\sigma_{\text{QoI},j}$ of QoIs $y_{ij}(i = 1 \dots 60)$ and $\sigma_{\text{output},j}$ of meteorological fields $M_{ij}^{kl}(i = 1 \dots 60, k = 1 \dots 50, l = 1 \dots 60)$ are used. It should be noted that other normalization methods are conceivable, which could, in turn, influence the weighting. This further highlights the issue with selecting fixed weights, thus motivating an investigation into the impact of different weights in this study.

Investigations in previous chapters have shown that three of the considered model parameters – namely the entrainment rate, the terminal fall speed of ice crystals, and the soil moisture evaporation fraction – exert a substantially stronger impact on the considered meteorological variables compared to the other three parameters (see Section 4.4.1, 4.4.2). Optimizing parameters with little or no impact on the objectives would not yield meaningful results. In particular, optimal values might be found at the boundaries of the allowed parameter range, despite having no notable effect on the objectives, a phenomenon that could be referred to as over-tuning. To avoid this issue, the optimization studies are restricted to the three parameters with the strongest influence on the considered meteorological variables. This approach also simplifies the interpretation of results by reducing the risk of over-tuning and ensuring that the parameter space remains more tractable. In the employed surrogate models, the other three parameters – which are not investigated here – are set to their default values. The parameter boundaries in the optimization process are set at the 1 % and 99 % levels of the respective CDFs. These limits are considered appropriate to maintain the physical plausibility of the parameter values while accounting for the reduced accuracy of surrogate models at the distribution tails, where the training point density is lower.

For both cases – the consideration of QoIs and meteorological fields – both optimization strategies proposed in Section 3.2 are applied. The sample size is chosen to balance result integrity and computational efficiency. For the *weight uncertainty* method, a total of 1000 samples is generated using MC simulation. In contrast, for the weight variation method, a total of 15 incremental steps is applied to adjust the weight of each objective. All optimization problems are solved using the SciPy package for Python, employing the Nelder-Mead method [137], which is well-suited because the objective function is treated as a black box, may be noisy, and has no directly available derivative with respect to the model parameters.

### 4.5.3 Optimization results

First, the optimization results based on QoIs are presented, followed by the consideration of meteorological fields. In both cases, the *weight uncertainty* method is applied first, followed by the *weight variation* method, to assess the impact of objective weighting on the optimal model parameters. The optimization yields results for uniform model parameters, which are then transformed back to the physical parameter space using PIT (Section 3.1.1) for graphical visualization.

**Optimization based on quantities of interest**

Figure 4.12 presents histograms of the three optimized model parameters obtained using the *weight uncertainty* method based on the QoIs. For reference, the original parameter PDFs are shown in light grey. For *entrorg*, the histogram collapses towards optimal values near the lower boundary defined in the optimization process. This result indicates that the optimization is primarily driven by the attempt to enhance rainfall in ICON, aligning it more closely with the wetter GPM IMERG data. The most effective way to achieve this is by reducing entrainment rates, pushing the optimal values toward the lowest plausible level. However, given the potential influence of topographic rainfall enhancement (see Section 4.5.1), this adjustment may improve agreement with reference data but does not necessarily result in a physically more realistic model configuration. In contrast, the other two parameters, *zvz0i* and *c_soil*, tend to converge toward values near the means of their original PDFs. Specifically, *zvz0i* is narrowly clustered slightly above its mean ($\sim$ 1.7 $\mathrm{m\,s^{-1}}$), whereas *c_soil* is more broadly distributed slightly below its mean ($\sim$ 0.8). This suggests that the default settings for these two parameters already provide a relatively balanced configuration for the considered system. The pronounced histogram peaks for *entrorg* and *zvz0i* indicate a relative insensitivity to weight variations. This is further supported by the GSA results (Section 4.4.1), where most QoIs exhibited strong sensitivity to these two parameters, implying that even small adjustments would significantly impact most QoIs.

Figure 4.13 presents histograms of the optimized QoIs, corresponding to the optimized parameters shown in Fig. 4.12. Horizontal lines indicate the values obtained from

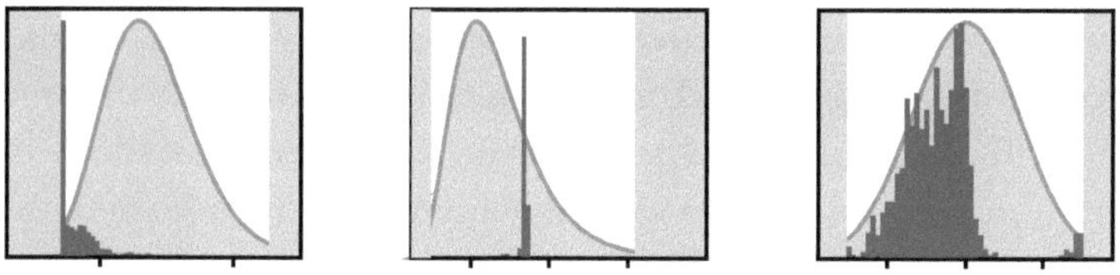


**Figure 4.12:** Histograms of optimal model parameters resulting from MC simulations using the weights from Table 4.6 for optimizing QoIs.

ICON simulations using default model parameters (solid) and reference data (dashed), corresponding to the triangles in Fig. 4.3. In general, QoIs with larger weights (Table 4.6) exhibit stronger shifts from their default values toward the reference values. The strong weighting for precipitation leads to an approximately 30 % improvement relative to GPM IMERG data. Similarly, 2 m dew point temperature and MSLP show significant improvements relative to ERA5 data, owing to their medium weights. However, column-integrated water vapor and 2 m temperature deteriorate, highlighting an inevitable trade-off in MOO when aiming for an overall optimum. The underlying cause of these changes is evident, as only the entrainment rate is significantly adjusted, and decreasing this parameter leads to increased precipitation, reduced 2 m temperature, higher 2 m dew point, a weaker SHL, and more column-integrated water vapor (see Fig. 4.3). Regarding cloud cover, only mid-level cloud cover exhibits a notable improvement, attributed to the slightly increased fall velocity of ice (Fig. 4.3). In contrast, low- and high-level cloud cover show no substantial improvement, primarily due to their lower weighting. As discussed in Section 4.5.1, accurately capturing low-level cloud cover remains a persistent challenge for both models and measurements, with none of the considered model parameters or their combinations yielding a fully satisfactory solution. This is further illustrated in Fig. 4.3, where for certain QoIs, such as low-level cloud cover, no parameter combination leads to alignment with the reference data. Additionally, QoIs with even lower weights exhibit diverse changes through the optimization process, including deterioration relative to the reference values (e. g., ITD latitude, AEJ and TEJ speeds and latitudes). This phenomenon illustrates that enhancing certain QoIs often compromises others, a characteristic indicative of results located on Pareto fronts. The inability to simultaneously optimize all QoIs to match reference data suggests potential physical inconsistencies either in the reference datasets or within the ICON model itself. The notable degradation of certain low-weighted QoIs further highlights the potential risk of overfitting to highly weighted QoIs.

Figure 4.14 presents the optimal model parameters for varied weights, where each data point represents a single-objective optimization result. The optimal parameters generally align with those in Fig. 4.12. However, substantial variations in the optimal parameters are observed for several QoIs when weights are adjusted. In some cases, despite clear overall trends, minor oscillations or jumps occur. These fluctuations are likely due to factors inherent in the optimization process, such as tolerances and numerical considerations, but could also indicate the existence of multiple local minima. Given the extensive methodology, which involves numerous separate optimization runs, a global optimization approach is considered computationally infeasible. However, these variations do not compromise the validity of the results, as the dominant trends remain well-defined. For high-weighted low-level cloud cover, significant parameter adjustments are observed, specifically higher entrainment rates and fall velocities of ice. This correlates with the substantially lower reference values of low-level clouds shown in Fig. 4.3. However, despite these adjustments, the achievable values for low-level cloud cover still deviate significantly from the reference values. As discussed in Section 4.5.1, enforcing an optimal match for this quantity remains challenging. For higher weights assigned to 2 m dew point temperature and accumulated precipitation, higher values of *c_soil* are preferred, as increased surface latent heat fluxes consistently lead to higher dew points. These findings are supported by Fig. 4.3, where higher *c_soil* values result in QoI values approaching the reference values. Regarding the considered latitudes, *entrorg* exhibits the strongest influence (see Fig. 4.3). Higher weights assigned to latitudes generally lead to larger optimal entrainment rates, which compress the latitudes and make the WAM system narrower, aligning it more closely with ERA5 reanalysis data. The relatively unpredictable changes in *c_soil*, which occasionally contradict expected trends based on the reference data in Fig. 4.3, can be attributed to its weaker influence compared to the other two parameters. As a result, the optimal parameters are predominantly determined by *entrorg* and *zvz0i*, with *c_soil* adjusting accordingly.

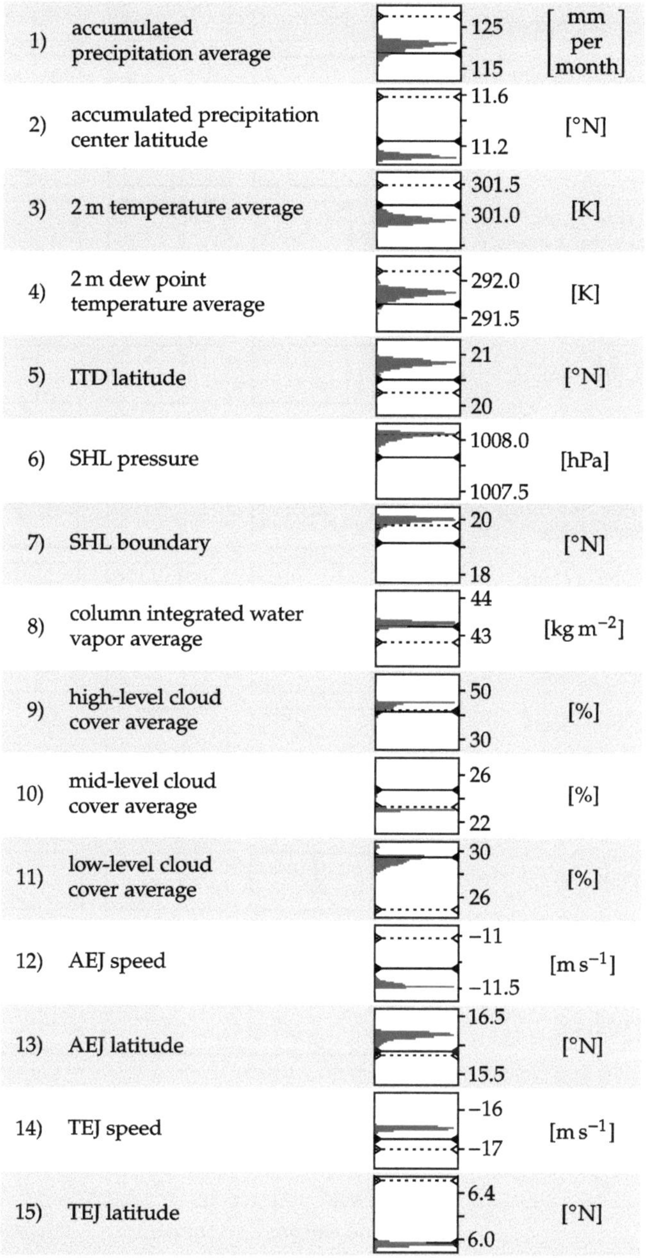


**Figure 4.13:** Histograms of QoIs corresponding to optimal model parameters (Fig. 4.12) obtained using the weights from Table 4.6 for optimizing QoIs. Horizontal lines indicate the values for ICON simulations with default model parameters (solid lines) and reference data (ERA5 and GPM IMERG; dashed lines).

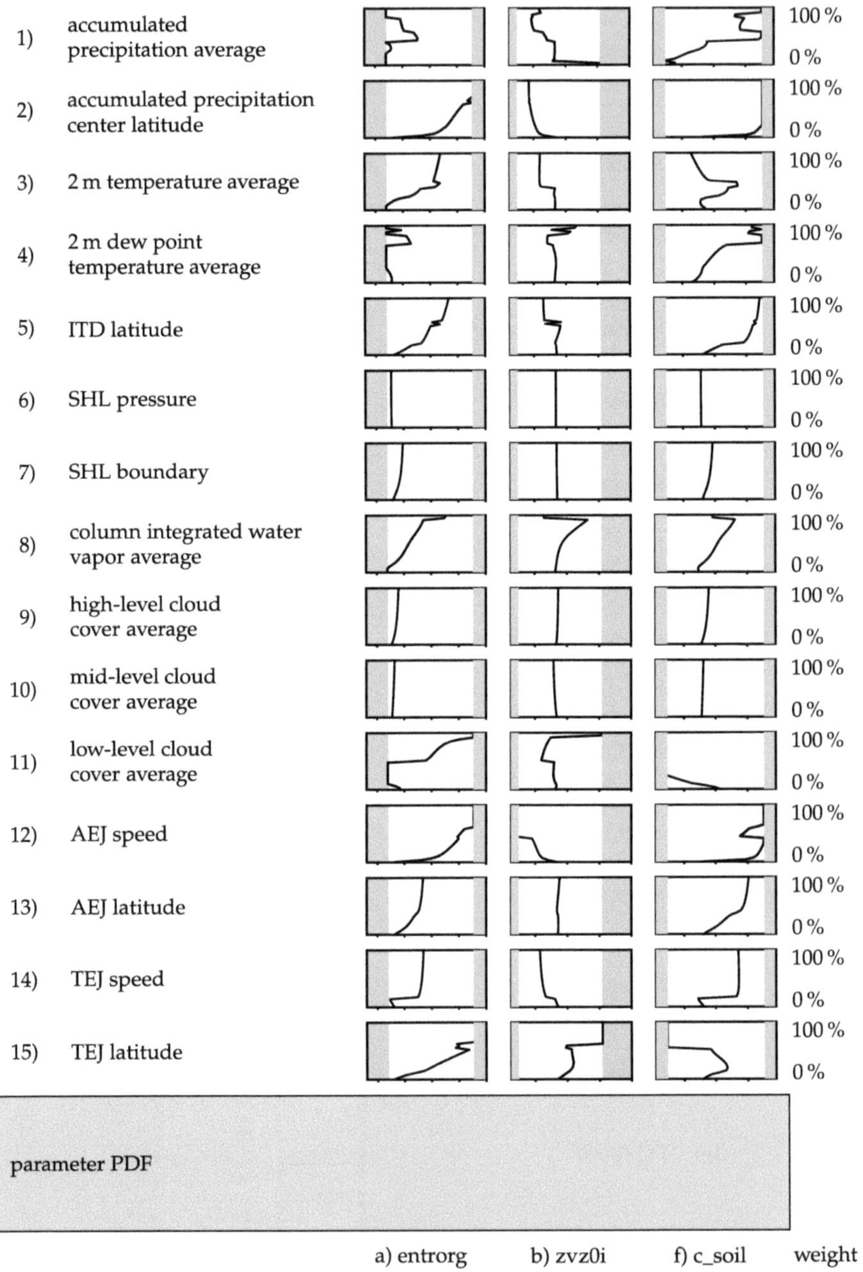


**Figure 4.14:** Optimal model parameters resulting from MOO. The weights (relative importance) for individual QoIs are successively increased, while the weights for all other QoIs are specified based on their default values according to Table 4.6, ensuring a total sum of 100 %. A separate optimization problem is solved for each QoI and weight combination.

### Optimization based on meteorological fields

In contrast to the previous section, this optimization is based on meteorological fields. Figure 4.15 presents histograms of the optimized model parameters obtained using the *weight uncertainty* method. The optimized parameters generally cluster near their default values, with slightly lower values for *entrorg*, slightly higher values for *zvz0i*, and lower values for *c_soil*. This suggests that the default parameter settings already provide a reasonable balance across all output variables for the specified time period and region, given the weight ranges from Table 4.6. The parameter values for *entrorg* and *zvz0i* exhibit greater concentration compared to *c_soil*, further highlighting their stronger influence on meteorological variables. This is supported by the larger magnitudes in the spatial variability fields (compare Fig. 4.5 with Figs. 4.6, 4.7). These variability fields, when compared with the target difference fields (Fig. 4.10), provide insights into whether modifying individual model parameters can achieve the desired changes in specific meteorological variables. For several variables, the spatial patterns of the spatial variability fields and the target difference fields differ significantly, particularly for precipitation, column-integrated water vapor, 2 m temperature, 2 m dew point temperature, MSLP, and, to a lesser extent, wind speeds. This suggests that adjusting model parameters alone would not lead to an overall improvement across the entire domain. For precipitation, parameter modifications primarily induce zonally oriented changes (Figs. 4.5a1, b1; 4.7f1), which do not align with the target differences, particularly over mountainous regions such as the Guinea Highlands and the Cameroon Line. This discrepancy is likely linked to the representation of convection, which is parameterized rather than explicitly resolved. While parameterized convection can yield realistic rainfall amounts – for example, over the DACCIWA region, as demonstrated by Kniffka, Knippertz, and Fink [63] – it still exhibits notable spatial discrepancies (Fig. 4.10.1). Additionally, the model's spatial resolution may struggle to capture complex topography and coastal dynamics. These discrepancies underscore the need for alternative approaches, such as increasing model resolution in simulations and improving physical representations within the model.

While the spatial variability fields for certain parameters exhibit some correlation with the target difference fields, the potential for achieving a combined optimal state across all variables remains limited. This limitation arises when the desired changes across multiple variables do not align with the directions of change induced by parameter adjustments. A notable example of this effect is 2 m temperature and 2 m dew point temperature. Increasing *c_soil* enhances evaporation, leading to lower low-level temperatures and higher dew point temperatures (see Fig. 4.3, opposite sign). However, since the differences between ICON and ERA5 are predominantly of the

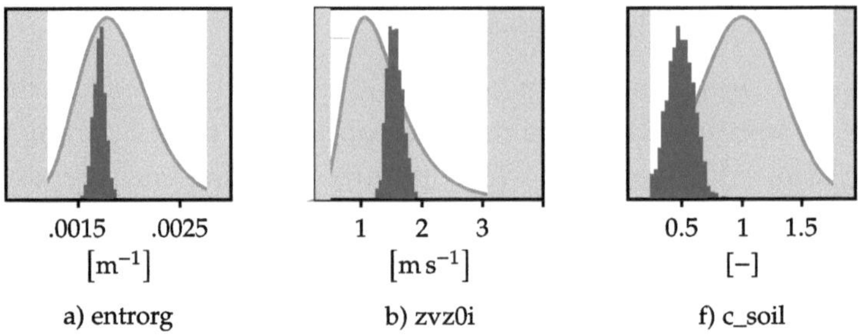


**Figure 4.15:** Histograms of optimal model parameters resulting from MC simulations using the weights from Table 4.6 for optimizing meteorological fields.

same sign (Fig. 4.10.2 and 3), adjusting *c_soil* cannot resolve this discrepancy. Thus, modifying the three selected parameters alone is insufficient for reducing the regional discrepancies between the ICON model outputs and the reference data.

Figure 4.16 presents the optimized parameters when varying the weights assigned to the meteorological variables, analogous to Fig. 4.14 for the QoI optimization. A notable difference compared to the QoI optimization is the increase in the optimal entrainment rate when higher weights are assigned to precipitation and high-level clouds. This suggests that a higher entrainment rate can more effectively address regional changes in these variables but not necessarily improve the domain-average values. For precipitation, in particular, the increase over the Guinea Highlands and the decrease over the eastern Gulf of Guinea under higher entrainment rates are dominant effects (Fig. 4.5a1), which guide the optimization towards the reference data (Fig. 4.10.1). Other dependencies of the optimal model parameters on assigned weights, as well as discrepancies between the two optimization strategies, provide valuable insights for model developers to refine parameter definitions or to better understand their effects. However, these findings should be interpreted with caution, given the limited domain and parameter set considered in this study.

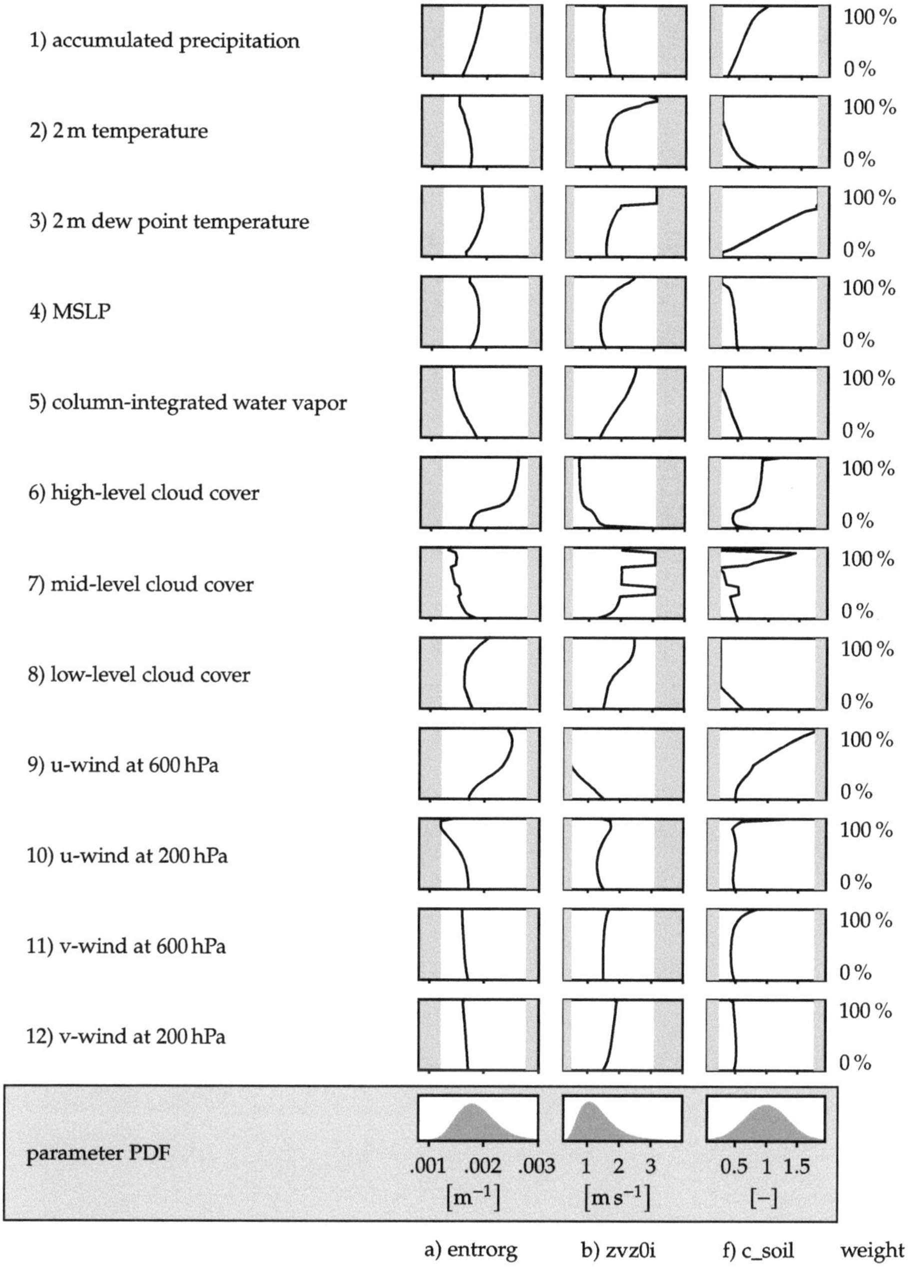


**Figure 4.16:** Optimal model parameters resulting from MOO. The weights (relative importance) (relative importance) for individual meteorological fields are successively increased, while weights for all other meteorological fields are specified based on their default values according to Table 4.6, ensuring a total sum of 100 %. A separate optimization problem is solved for each meteorological field and weight combination.

# 5 Conclusions

This section first summarizes the key findings of the study. It then discusses limitations and perspectives, providing a foundation for future research and further investigations.

## 5.1 Summary

The aim of this study was to develop surrogate models as cost-effective alternatives to represent the characteristics of the WAM system during boreal summer, simulated using the ICON model. These surrogate models were designed to explore how model parameters contribute to uncertainties and to conduct parameter optimization studies using reference data. To achieve these objectives, methods related to surrogate modeling, UQ, and optimization were carefully selected, with methodological improvements made as needed for effectiveness. In particular, surrogate models using PCR and GPR were constructed to represent meteorological fields and scalar QoIs.

The investigations in this work lead to the following conclusions:

- In UQ, a comprehensive framework that combines space-filling designs in a uniform input space with input space transformation in surrogate models has proven to be an effective, streamlined, and accurate approach. This method removes the need to set arbitrary bounds at the tails of PDFs in the parameter space. Additionally, it aligns the density of training points with the PDF values, making such methods more accurate in regions that are sampled more frequently, which is particularly beneficial for applications like GSA. Overall, this method combines the advantages of space-filling designs, such as LHS, which are particularly suited for a unit uniform input space, with the benefits of considering the physical input space – directly related to the computational model – when defining trend functions that capture general physical tendencies.
- In MOO, variations in objective weights can significantly impact optimization outcomes. In the meteorological study, parameters with high uncertainty contri-

butions to outputs were less sensitive to changes in objective weights compared to those with lower contributions. This finding highlights a potential risk: parameters deemed less critical, with weaker effects on system outputs, may respond strongly to weight variations. This increases the risk of overfitting, especially in systems with uncertainties where unmodeled processes prevent any parameter combination from fully matching reference data. While other studies that aimed to reduce the multi-objective problem using similar weighted-sum approaches typically relied on a fixed combination of weights – often assigning equal weights to all variables – to keep the dimensionality manageable for both computation and interpretation, the approach proposed in this work offers a more flexible and generalized framework. By allowing for systematic weight variations, it provides higher evidence for the optimization results.

- In the meteorological context, the WAM system exemplifies a complex, interdependent system where model parameters can have counterintuitive effects. Key QoIs are influenced by multiple parameters, reflecting the coupled relationships within the WAM system. Three of the considered uncertain parameters strongly influence WAM characteristics: the entrainment rate, terminal fall speed of ice crystals and soil moisture evaporation fraction. In particular, an increase in the first two – both directly affecting cloudy regions and convective processes – leads to a substantial reduction in cloud cover and precipitation, except for a distinct increase along a narrow strip in the center of the rain belt. Overall, these changes contribute to warmer and drier conditions over the Sahara.
- Results indicate that model parameters in the default setup are generally well-calibrated, although optimal values are sensitive to the weighting of meteorological variables in the optimization. For instance, a low entrainment rate improves the representation of average WAM quantities, including increased rainfall, higher 2 m dew points, and a weaker SHL. For meteorological fields, default parameters provide a reasonable balance. In most cases, parameter changes alter patterns without achieving desired spatial alignment with reference data, suggesting that these parameters cannot fully account for spatial discrepancies. This indicates that the underlying spatial mismatches likely originate from structural issues within the model rather than parameterization choices. Furthermore, improving accuracy for specific variables often compromises others.
- Supported by an interactive tool, the findings provide meteorologists with valuable insights for refining parameter specifications and improving model performance. In the context of the WAM system, the use of PCR-based surrogates to illustrate the variability of spatial fields offers a novel approach for analyzing parameter impacts.

## 5.2 Limitations and Perspective

In this study, the analysis was based on climatological mean values over a four-year period, disregarding interannual trends. While averaging over multiple years mitigates the influence of year-to-year variability, some dependence on the selected years may still persist. Consequently, the results do not represent instantaneous WAM characteristics for a specific forecast time but instead provide insights into its mean state. For these findings to contribute to improving daily weather predictions, further investigations would be necessary, particularly focusing on individual forecasts and their diurnal variability. This could include statistical analyses to assess the temporal and spatial spread of forecast variables, providing deeper insights into short-term predictability and potential forecast improvements.

Regarding the selected uncertain model parameters (Section 4.1.2), given the limited information available in the literature, these definitions are rough estimates, and the results should be interpreted with caution. Particularly, the results in GSA are directly influenced by the parameter PDFs, where changes in PDF spread can significantly impact the sensitivity indices.

The findings of this study suggest that the entrainment rate, ice fall speed, and surface evaporation parameters have the stronges impact on monsoon characteristics and therefore require more precise specification to further enhance model accuracy in the considered region. This could be achieved through more and different types of measurements, more comprehensive expert input or improvement of physical representation of processes related to these parameters. Furthermore, this study has shown that some parameters exert only a marginal influence on the considered system, yet they were included in the entire surrogate modeling procedure. This suggests that, for future studies, it may be advisable to first conduct one-at-a-time studies to assess whether a parameter has a significant impact and should be included in the analysis or whether the parameter space could be reduced to enhance computational efficiency and focus on the most relevant parameters.

Additionally, only six model parameters were included in the parameter studies and only the three parameters with strongest impact were considered for the optimization studies. However, other parameters or processes may also contribute relevant uncertainties. This limited selection risks overfitting, as the optimization may adjust model parameters to values that improve alignment of the ICON model output with reference data, even when discrepancies are caused by processes not captured by these parameters. The limitation of the optimization to three parameters reduces the risk of overtuning the parameters with only weak impacts, however, it also increases the risk of mistuning the parameters with strong impacts. Furthermore, results from parameter optimization studies - even though parameters may have a similar effect in other parts of the atmosphere - are specific to the West African region. Consequently, parameters optimized for specific regions might be implemented in the weather model to be only valid within those areas, similar to *rhebc_land_trop* and *rcucov_trop*, which are tuned for tropical regions.

Results based on this version of the ICON model cannot necessarily be generalized to other model versions or models, which may use different parameters in their parameterizations. Nevertheless, this study demonstrates the utility of the applied methodology, including the training procedure and surrogate models. The methodology is adaptable and can accommodate more parameters, with computational costs expected to increase linearly with the number of parameters. Given the strong influence of the entrainment rate and ice fall speed observed in this study, other deep-cloud-related parameters may also be worthwhile to investigate, with cloud inhomogeneity being one particularly interesting aspect for future studies.

Moreover, this study highlights the limitations of parameter tuning in meteorology, particularly in regions influenced by factors like complex topography or coastal dynamics. To further enhance model accuracy, alternative strategies should be explored, such as increasing spatial resolution, refining physical process representations, or adopting novel approaches like artificial intelligence for weather prediction and climate projection [4].

From a methodological perspective, the study could be enhanced by reevaluating the surrogate modeling methods and comparing alternative approaches. For instance, rather than using separate models for meteorological fields and scalar QoIs, a unified framework could be developed. This could involve calculating QoIs directly from surrogate models for meteorological fields, though this approach would increase computational demands, as each sample from the surrogate model would require recalculating the QoIs from the fields. Alternatively, to improve surrogate models for field data by means of PCR, GPR could be applied to the PCA scores rather than relying on linear (or transformed linear) models. Other surrogate modeling techniques, such as polynomial chaos expansion, artificial neural networks, and support vector regression, are less represented in meteorological literature but may offer valuable insights, though they often require more training data and may lack uncertainty quantification. Comparative studies between these approaches could highlight the strengths and limitations of each method in specific contexts.

The proposed methodology, which employs transformed input parameter spaces, has demonstrated its relevance in the context of UQ. By increasing the density of training points in regions of the input space associated with higher probabilities, this approach aligns naturally with GSA, where more samples are drawn in regions of high probability density. However, if the primary objective were to develop a surrogate model with uniform accuracy across the entire parameter space – including the tails of the probability distributions – a uniform training point density would be more appropriate. In such a case, the choice of parameter space boundaries becomes crucial, as probability distributions may not be strictly bounded. Setting overly wide bounds would result in a large proportion of training points being drawn from the distribution tails, reducing surrogate model accuracy in regions of high probability. Conversely, choosing narrow bounds would confine the surrogate model to a limited region, rendering it inapplicable beyond these bounds. A reasonable compromise is to define the parameter space using the $3\sigma$-bounds of the distributions. The framework presented in this study addresses this decision challenge by transforming the input space to a uniform distribution, thereby ensuring a systematic and adaptive approach to training point allocation.

A natural extension of parameter optimization might be the use of Bayesian methods to update prior distributions based on the optimized parameter values. However, since the prior distributions in this study are only rough estimates without a solid foundational basis, and since optimization inherently incorporates effects that may lead to implausible values – such as over-tuning, incorrect tuning due to missing model parameters or processes, or limitations imposed by the domain of the framework – combining these datasets would not be appropriate. Instead, the focus should be on analyzing the optimal parameter values identified for this specific case in comparison to the assumed default parameter values and ranges, which are also utilized by the DWD in their ensemble forecasts.

If, however, the prior parameter distributions were derived from well-founded knowledge, and if the optimization results were obtained in a broader context – potentially including a global atmospheric model with a sufficiently comprehensive representation of parameters and processes – then one could assume that optimization-based tuning would yield reliable and globally applicable results. In such a case, Bayesian updating of prior probability distributions using optimization results could be a meaningful and promising next step.

Considering multiple outputs for both QoIs and meteorological fields, a useful extension beyond separate models may be to employ comprehensive multi-output regression models. For scalar quantities, co-kriging is a common extension of GPR for handling multiple outputs. Other multi-output regression methods, such as multi-output support vector regression [150], have been shown to improve prediction accuracy when outputs are correlated, though mainly in simple examples. In preliminary studies for this work and a Master's thesis [162], multi-output methods did not yield significant accuracy improvements and were deemed computationally prohibitive due to their high costs relative to the modest, if any, gains in accuracy.

# Appendix

# A Matrix identities

Derivatives of the elements of an inverse matrix:

$$\frac{\partial}{\partial x}\mathbf{M}(x)^{-1} = -\mathbf{M}(x)^{-1}\frac{\partial \mathbf{M}(x)}{\partial x}\mathbf{M}(x)^{-1}$$

Derivative of the log determinant of a positive definite symmetric matrix:

$$\frac{\partial}{\partial x}\log(|\mathbf{M}(x)|) = \mathrm{tr}\left(\mathbf{M}(x)^{-1}\frac{\partial \mathbf{M}(x)}{\partial x}\right)$$

Cyclic permutation of matrices in the argument of a trace:

$$\mathrm{tr}(\mathbf{M}_1\,\mathbf{M}_2\,\mathbf{M}_3) = \mathrm{tr}(\mathbf{M}_2\,\mathbf{M}_3\,\mathbf{M}_1) = \mathrm{tr}(\mathbf{M}_3\,\mathbf{M}_1\,\mathbf{M}_2) \tag{A.1}$$

# B Gradient of the log marginal likelihood for universal kriging

Hyperparameter optimization can be computationally intensive for high-dimensional problems with a large number of training points. For simple kriging, the gradient of the log marginal likelihood has been shown in Eq. 2.3. In the following, the derivation of the gradient of the log marginal likelihood for universal kriging is carried out as shown by Fischer and Proppe [158], where suitable abbreviations for efficient computation are introduced.

The log marginal likelihood for universal kriging, as shown by Rasmussen and Williams [113], is

$$\log p(\mathbf{y}|\mathbf{U},\boldsymbol{\theta}) = -\frac{1}{2}\mathbf{y}^\top\mathbf{K}^{-1}\mathbf{y} + \frac{1}{2}\mathbf{y}^\top\mathbf{C}\mathbf{y} - \frac{1}{2}\log|\mathbf{K}| - \frac{1}{2}\log|\mathbf{A}| - \frac{n-m}{2}\log 2\pi\,, \tag{B.1}$$

where $\mathbf{A} = \mathbf{H}\mathbf{K}^{-1}\mathbf{H}^\top$, $\mathbf{C} = \mathbf{K}^{-1}\mathbf{H}^\top\mathbf{A}^{-1}\mathbf{H}\mathbf{K}^{-1}$ and $m = \mathrm{rank}(\mathbf{H}^\top)$.
Inserting $\mathbf{A}$ and $\mathbf{C}$ in B.1 results in

$$\begin{aligned}\log p(\mathbf{y}|\mathbf{U},\boldsymbol{\theta}) =& -\frac{1}{2}\mathbf{y}^\top\mathbf{K}^{-1}\mathbf{y} + \frac{1}{2}\mathbf{y}^\top\left(\mathbf{K}^{-1}\mathbf{H}^\top(\mathbf{H}\mathbf{K}^{-1}\mathbf{H}^\top)^{-1}\mathbf{H}\mathbf{K}^{-1}\right)\mathbf{y}\\ &-\frac{1}{2}\log|\mathbf{K}| - \frac{1}{2}\log|\mathbf{H}\mathbf{K}^{-1}\mathbf{H}^\top| - \frac{n-m}{2}\log 2\pi\\ =& -\frac{1}{2}\mathbf{y}^\top\boldsymbol{\alpha} + \frac{1}{2}\mathbf{y}^\top\boldsymbol{\gamma}\boldsymbol{\eta}\boldsymbol{\alpha}\\ &-\frac{1}{2}\log|\mathbf{K}| - \frac{1}{2}\log|\mathbf{H}\boldsymbol{\gamma}| - \frac{n-m}{2}\log 2\pi,\end{aligned}$$

where $\boldsymbol{\alpha}$, $\boldsymbol{\gamma}$ and $\boldsymbol{\eta}$ are defined as

$$\boldsymbol{\alpha} = \mathbf{K}^{-1}\mathbf{y} = \mathbf{L}_K^{-\top}(\mathbf{L}_K^{-1}\mathbf{y})\,, \tag{B.2}$$

$$\boldsymbol{\gamma} = \mathbf{K}^{-1}\mathbf{H}^\top = \mathbf{L}_K^{-\top}(\mathbf{L}_K^{-1}\mathbf{H}^\top)\,, \tag{B.3}$$

$$\boldsymbol{\eta} = (\mathbf{H}\mathbf{K}^{-1}\mathbf{H}^\top)^{-1}\mathbf{H} = \mathbf{L}_\eta^{-\top}(\mathbf{L}_\eta^{-1}\mathbf{H}) \tag{B.4}$$

with $\mathbf{L}_K = \text{cholesky}(\mathbf{K})$ and $\mathbf{L}_\eta = \text{cholesky}(\mathbf{H}\mathbf{K}^{-1}\mathbf{H}^\top)$. Cholesky decomposition is applied in order to efficiently determine the inverse.

Taking the derivative with respect to $\boldsymbol{\theta}$ by the use of matrix identities in Appendix A results in

$$\begin{aligned}
&\frac{\partial}{\partial\theta_l}\log p(\mathbf{y}|\mathbf{U},\boldsymbol{\theta})\\
&= \frac{1}{2}\mathbf{y}^\top\mathbf{K}^{-1}\frac{\partial\mathbf{K}}{\partial\theta_l}\mathbf{K}^{-1}\mathbf{y}\\
&-\frac{1}{2}\mathbf{y}^\top\left(\mathbf{K}^{-1}\frac{\partial\mathbf{K}}{\partial\theta_l}\mathbf{K}^{-1}\mathbf{H}^\top(\mathbf{H}\mathbf{K}^{-1}\mathbf{H}^\top)^{-1}\mathbf{H}\mathbf{K}^{-1}\right)\mathbf{y}\\
&+\frac{1}{2}\mathbf{y}^\top\left(\mathbf{K}^{-1}\mathbf{H}^\top(\mathbf{H}\mathbf{K}^{-1}\mathbf{H}^\top)^{-1}\mathbf{H}\mathbf{K}^{-1}\frac{\partial\mathbf{K}}{\partial\theta_l}\mathbf{K}^{-1}\mathbf{H}^\top(\mathbf{H}\mathbf{K}^{-1}\mathbf{H}^\top)^{-1}\mathbf{H}\mathbf{K}^{-1}\right)\mathbf{y}\\
&-\frac{1}{2}\mathbf{y}^\top\left(\mathbf{K}^{-1}\mathbf{H}^\top(\mathbf{H}\mathbf{K}^{-1}\mathbf{H}^\top)^{-1}\mathbf{H}\mathbf{K}^{-1}\frac{\partial\mathbf{K}}{\partial\theta_l}\mathbf{K}^{-1}\right)\mathbf{y}\\
&-\frac{1}{2}\text{tr}\left(\mathbf{K}^{-1}\frac{\partial\mathbf{K}}{\partial\theta_l}\right)-\frac{1}{2}\text{tr}\left(-(\mathbf{H}\mathbf{K}^{-1}\mathbf{H}^\top)^{-1}\mathbf{H}\mathbf{K}^{-1}\frac{\partial\mathbf{K}}{\partial\theta_l}\mathbf{K}^{-1}\mathbf{H}^\top\right).
\end{aligned}$$

In the next step, symmetry and positive definiteness of $\mathbf{K}$ (by definition) and therefore symmetry of $\mathbf{K}^{-1}$ are taken into account. Furthermore, the matrix identity in Eq. A.1 is used to simplify the result. With $\boldsymbol{\alpha}, \boldsymbol{\eta}$ and $\boldsymbol{\gamma}$, it follows

$$\begin{aligned}
&\frac{\partial}{\partial\theta_l}\log p(\mathbf{y}|\mathbf{U},\boldsymbol{\theta})\\
&= \frac{1}{2}\boldsymbol{\alpha}^\top\frac{\partial\mathbf{K}}{\partial\theta_l}\boldsymbol{\alpha}-\frac{1}{2}\boldsymbol{\alpha}^\top\frac{\partial\mathbf{K}}{\partial\theta_l}\boldsymbol{\gamma}\boldsymbol{\eta}\boldsymbol{\alpha}+\frac{1}{2}\boldsymbol{\alpha}^\top\boldsymbol{\eta}^\top\boldsymbol{\gamma}^\top\frac{\partial\mathbf{K}}{\partial\theta_l}\boldsymbol{\gamma}\boldsymbol{\eta}\boldsymbol{\alpha}\\
&-\frac{1}{2}\boldsymbol{\alpha}^\top\boldsymbol{\eta}^\top\boldsymbol{\gamma}^\top\frac{\partial\mathbf{K}}{\partial\theta_l}\boldsymbol{\alpha}-\frac{1}{2}\text{tr}\left(\mathbf{K}^{-1}\frac{\partial\mathbf{K}}{\partial\theta_l}\right)+\frac{1}{2}\text{tr}\left(\boldsymbol{\gamma}\boldsymbol{\eta}\mathbf{K}^{-1}\frac{\partial\mathbf{K}}{\partial\theta_l}\right)\\
&= \frac{1}{2}\text{tr}\left(\boldsymbol{\alpha}^\top\frac{\partial\mathbf{K}}{\partial\theta_l}\boldsymbol{\alpha}-\boldsymbol{\alpha}^\top\frac{\partial\mathbf{K}}{\partial\theta_l}\boldsymbol{\gamma}\boldsymbol{\eta}\boldsymbol{\alpha}+\boldsymbol{\alpha}^\top\boldsymbol{\eta}^\top\boldsymbol{\gamma}^\top\frac{\partial\mathbf{K}}{\partial\theta_l}\boldsymbol{\gamma}\boldsymbol{\eta}\boldsymbol{\alpha}-\boldsymbol{\alpha}^\top\boldsymbol{\eta}^\top\boldsymbol{\gamma}^\top\frac{\partial\mathbf{K}}{\partial\theta_l}\boldsymbol{\alpha}\right)\\
&-\frac{1}{2}\text{tr}\left(\mathbf{K}^{-1}\frac{\partial\mathbf{K}}{\partial\theta_l}\right)+\frac{1}{2}\text{tr}\left(\boldsymbol{\gamma}\boldsymbol{\eta}\mathbf{K}^{-1}\frac{\partial\mathbf{K}}{\partial\theta_l}\right)\\
&= \frac{1}{2}\text{tr}\left(\left(\boldsymbol{\alpha}\boldsymbol{\alpha}^\top-\boldsymbol{\gamma}\boldsymbol{\eta}\boldsymbol{\alpha}\boldsymbol{\alpha}^\top+\boldsymbol{\gamma}\boldsymbol{\eta}\boldsymbol{\alpha}\boldsymbol{\alpha}^\top\boldsymbol{\eta}^\top\boldsymbol{\gamma}^\top-\boldsymbol{\alpha}\boldsymbol{\alpha}^\top\boldsymbol{\eta}^\top\boldsymbol{\gamma}^\top-\mathbf{K}^{-1}+\boldsymbol{\gamma}\boldsymbol{\eta}\mathbf{K}^{-1}\right)\frac{\partial\mathbf{K}}{\partial\theta_l}\right).
\end{aligned}$$

The abbreviations $\boldsymbol{\rho} = \boldsymbol{\alpha}\boldsymbol{\alpha}^\top$, $\boldsymbol{\varepsilon} = \boldsymbol{\gamma}\boldsymbol{\eta}$ and $\boldsymbol{\xi} = \boldsymbol{\varepsilon}\boldsymbol{\rho}$ are introduced. It follows

$$\frac{\partial}{\partial\theta_l}\log p(\mathbf{y}|\mathbf{U},\boldsymbol{\theta}) = \frac{1}{2}\text{tr}\left(\left(\boldsymbol{\rho}-\boldsymbol{\xi}-\boldsymbol{\xi}^\top+\boldsymbol{\xi}\boldsymbol{\varepsilon}^\top+(\boldsymbol{\varepsilon}-\mathbb{1})\mathbf{K}^{-1}\right)\frac{\partial\mathbf{K}}{\partial\theta_l}\right).$$

Using the Einstein summation convention, the following calculation steps are required to determine the derivative of the log marginal likelihood given $\boldsymbol{\alpha}$, $\gamma$ and $\eta$ from Eq. B.2, Eq. B.3 and Eq. B.4:

$$
\begin{aligned}
\rho_{ij} &= \alpha_i \alpha_j \,, \\
\varepsilon_{ij} &= \gamma_{ik} \eta_{kj} \,, \\
\xi_{il} &= \varepsilon_{ij} \rho_{jl} \,, \\
\frac{\partial}{\partial \theta_l} &\log p(\mathbf{y}|\mathbf{U}, \boldsymbol{\theta}) \\
&= \frac{1}{2} \left( \left( \rho_{ij} - \xi_{ij} - \xi_{ji} + \xi_{im} \varepsilon_{jm} + (\varepsilon_{im} - \delta_{im}) \left[ \mathbf{K}^{-1} \right]_{mj} \right) \frac{\partial K_{ji}}{\partial \theta_l} \right)
\end{aligned} \tag{B.5}
$$

with Kronecker delta $\delta_{im}$.

The complexity of the computation of the gradient of the log marginal likelihood (Eq. 2.3 and Eq. B.5) is dominated by the inverse of matrix $\mathbf{K}$ which is of the computational complexity $\mathcal{O}(n^3)$. Once the inverse is determined, it can be used for the computation of all hyperparameters $\theta_l$. In contrast, computation of the gradient of the log marginal likelihood based on Eq. B.1, i. e. without using the gradient of matrix $\mathbf{K}$, is of the computational complexity $\mathcal{O}(p \cdot n^3)$, because the inverse of $\mathbf{K}$ has to be determined towards all input dimensions $p$. Gradient-based hyperparameter optimization with incorporation of gradient information is therefore more beneficial.

# List of Abbreviations

| | |
|---|---|
| **AEJ** | African Easterly Jet |
| **CDF** | cumulative distribution function |
| **DWD** | Deutscher Wetterdienst (German Meteorological Service) |
| **ERA5** | reanalysis data provided by the European Centre for Medium-Range Weather Forecasts |
| **FAST** | Fourier Amplitude Sensitivity Test |
| **GPM** | Global Precipitation Measurement |
| **GPR** | Gaussian process regression |
| **GSA** | global sensitivity analysis |
| **i. i. d.** | independent and identically distributed |
| **ICON** | Icosahedral Non-Hydrostatic (weather forecasting model) |
| **IMERG** | Integrated Multi-satellitE Retrievals for GPM |
| **ITD** | Intertropical Discontinuity |
| **LHD** | Latin hypercube design |
| **LHS** | Latin hypercube sampling |
| **MC** | Monte Carlo |
| **MOO** | multi-objective optimization |
| **MSE** | mean square error |
| **MSLP** | mean sea-level pressure |
| **NMSE** | normalized mean square error |
| **PCA** | principal component analysis |
| **PCR** | principal component regression |
| **PDF** | probability density function |
| **PIT** | probability integral transformation |
| **PPF** | percent-point function |
| **QoI** | quantity of interest |
| **RMSE** | root mean square error |
| **SHL** | Saharan heat low |
| **SST** | sea surface temperature |

| | |
|---|---|
| **TEJ** | Tropical Easterly Jet |
| **UQ** | uncertainty quantification |
| **WAM** | West African monsoon |

# List of Figures

# List of Tables

# Bibliography


[1] Agustí-Panareda, A., A. Beljaars, C. Cardinali, I. Genkova, and C. Thorncroft (2010): Impacts of Assimilating AMMA Soundings on ECMWF Analyses and Forecasts. *Weather and Forecasting* **25**(4), pp. 1142–1160. doi: 10.1175/2010waf2222370.1.

[2] Babaei, M., A. Alkhatib, and I. Pan (2015): Robust optimization of subsurface flow using polynomial chaos and response surface surrogates. *Computational Geosciences* **19**(5), pp. 979–998. doi: 10.1007/s10596-015-9516-5.

[3] Bellprat, O., S. Kotlarski, D. Lüthi, and C. Schär (2012): Objective calibration of regional climate models. *Journal of Geophysical Research: Atmospheres* **117**(D23), p. 23115. doi: 10.1029/2012jd018262.

[4] Ben Bouallègue, Z., M. C. A. Clare, L. Magnusson, E. Gascón, M. Maier-Gerber, M. Janoušek, M. Rodwell, F. Pinault, J. S. Dramsch, S. T. K. Lang, B. Raoult, F. Rabier, M. Chevallier, I. Sandu, P. Dueben, M. Chantry, and F. Pappenberger (2024): The Rise of Data-Driven Weather Forecasting: A First Statistical Assessment of Machine Learning–Based Weather Forecasts in an Operational-Like Context. *Bulletin of the American Meteorological Society* **105**(6), E864–E883. doi: 10.1175/bams-d-23-0162.1.

[5] Borchani, H., G. Varando, C. Bielza, and P. Larrañaga (2015): A survey on multi-output regression. *WIREs Data Mining and Knowledge Discovery* **5**(5), pp. 216–233. doi: 10.1002/widm.1157.

[6] Box, G. E. P. and P. W. Tidwell (1962): Transformation of the Independent Variables. *Technometrics* **4**(4), pp. 531–550. doi: 10.2307/1266288.

[7] Brierley, C. M., A. Zhao, S. P. Harrison, P. Braconnot, C. J. R. Williams, D. J. R. Thornalley, X. Shi, J.-Y. Peterschmitt, R. Ohgaito, D. S. Kaufman, M. Kageyama, J. C. Hargreaves, M. P. Erb, J. Emile-Geay, R. D'Agostino, D. Chandan, M. Carré, P. J. Bartlein, W. Zheng, Z. Zhang, Q. Zhang, H. Yang, E. M. Volodin, R. A. Tomas, C. Routson, W. R. Peltier, B. Otto-Bliesner, P. A. Morozova, N. P. McKay, G. Lohmann, A. N. Legrande, C. Guo, J. Cao, E. Brady, J. D. Annan, and A. Abe-Ouchi (2020): Large-scale features and

evaluation of the PMIP4-CMIP6 midHolocene simulations. *Climate of the Past* **16**(5), pp. 1847–1872. doi: 10.5194/cp-16-1847-2020.

[8] Burpee, R. W. (1972): The Origin and Structure of Easterly Waves in the Lower Troposphere of North Africa. *Journal of the Atmospheric Sciences* **29**(1), pp. 77–90. doi: 10.1175/1520-0469(1972)029<0077:toasoe>2.0.co;2.

[9] Chang, W., P. J. Applegate, M. Haran, and K. Keller (2014): Probabilistic calibration of a Greenland Ice Sheet model using spatially resolved synthetic observations: toward projections of ice mass loss with uncertainties. *Geoscientific Model Development* **7**(5), pp. 1933–1943. doi: 10.5194/gmd-7-1933-2014.

[10] Cinquegrana, D., A. L. Zollo, M. Montesarchio, and E. Bucchignani (2023): A Metamodel-Based Optimization of Physical Parameters of High Resolution NWP ICON-LAM over Southern Italy. *Atmosphere* **14**(5), p. 788. doi: 10.3390/atmos14050788.

[11] Claussen, M., A. Dallmeyer, and J. Bader (2017): Theory and Modeling of the African Humid Period and the Green Sahara. In: *Oxford Research Encyclopedia of Climate Science*. Oxford University Press. doi: 10.1093/acrefore/9780190228620.013.532.

[12] Cook, K. H. and E. K. Vizy (2006): Coupled Model Simulations of the West African Monsoon System: Twentieth- and Twenty-First-Century Simulations. *Journal of Climate* **19**(15), pp. 3681–3703. doi: 10.1175/jcli3814.1.

[13] Couvreux, F., F. Hourdin, D. Williamson, R. Roehrig, V. Volodina, N. Villefranque, C. Rio, O. Audouin, J. Salter, E. Bazile, F. Brient, F. Favot, R. Honnert, M.-P. Lefebvre, J.-B. Madeleine, Q. Rodier, and W. Xu (2021): Process-Based Climate Model Development Harnessing Machine Learning: I. A Calibration Tool for Parameterization Improvement. *Journal of Advances in Modeling Earth Systems* **13**, e2020MS002217. doi: 10.1029/2020MS002217.

[14] Dashti, M. and A. M. Stuart (2017): The Bayesian Approach to Inverse Problems. In: *Handbook of Uncertainty Quantification*. R. Ghanem and H. Higdon David and Owhadi eds. Cham: Springer International Publishing, pp. 311–428. doi: 10.1007/978-3-319-12385-1_7.

[15] Deutscher Wetterdienst (DWD) (2019): *ICON Namelist Overview*. Tech. rep.

[16] Devroye, L. (1986): *Non-Uniform Random Variate Generation*. Springer New York. doi: 10.1007/978-1-4613-8643-8.

[17] Diamond, M. S., H. M. Director, R. Eastman, A. Possner, and R. Wood (2020): Substantial Cloud Brightening From Shipping in Subtropical Low Clouds. *AGU Advances* **1**(1). doi: 10.1029/2019av000111.

[18] Dubrule, O. (1983): Cross validation of kriging in a unique neighborhood. *Journal of the International Association for Mathematical Geology* **15**, pp. 687–699. doi: 10.1007/bf01033232.

[19] Efron, B., T. Hastie, I. Johnstone, and R. Tibshirani (2004): Least angle regression. *The Annals of Statistics* **32**(2), pp. 407–499. doi: 10.1214/009053604000000067.

[20] Eldred, M. S., H. Agarwal, V. M. Perez, S. F. Wojtkiewicz Jr., and J. E. Renaud (2007): Investigation of reliability method formulations in DAKOTA/UQ. *Structure and Infrastructure Engineering* **3**(3), pp. 199–213. doi: 10.1080/15732470500254618.

[21] Eldred, M., C. Webster, and P. Constantine (2008): Evaluation of Non-Intrusive Approaches for Wiener-Askey Generalized Polynomial Chaos. *49th AIAA/ASME/ASCE/AHS/ASC Structures, Structural Dynamics, and Materials Conference*. doi: 10.2514/6.2008-1892.

[22] Fink, A. H. and A. Reiner (2003): Spatiotemporal variability of the relation between African easterly waves and West African squall lines in 1998 and 1999. *Journal of Geophysical Research* **108**(D11). doi: 10.1029/2002jd002816.

[23] Fink, A. H., A. Agustí-Panareda, D. J. Parker, J.-P. Lafore, J.-B. Ngamini, E. Afiesimama, A. Beljaars, O. Bock, M. Christoph, F. Didé, C. Faccani, N. Fourrié, F. Karbou, J. Polcher, Z. Mumba, M. Nuret, S. Pohle, F. Rabier, A. M. Tompkins, and G. Wilson (2011): Operational meteorology in West Africa: observational networks, weather analysis and forecasting. *Atmos. Sci. Lett.* **12**(1), pp. 135–141. doi: 10.1002/asl.324.

[24] Fink, A. H., T. Engel, V. Ermert, R. van der Linden, M. Schneidewind, R. Redl, E. Afiesimama, W. M. Thiaw, C. Yorke, M. Evans, and S. Janicot (2017): Mean Climate and Seasonal Cycle. In: *Meteorology of Tropical West Africa*. John Wiley & Sons, Ltd, pp. 1–39. doi: 10.1002/9781118391297.ch1.

[25] Flaounas, E., S. Bastin, and S. Janicot (2011): Regional climate modelling of the 2006 West African monsoon: sensitivity to convection and planetary boundary layer parameterisation using WRF. *Climate Dynamics* **36**(5-6), pp. 1083–1105. doi: 10.1007/s00382-010-0785-3.

[26] Fletcher, C. G., B. Kravitz, and B. Badawy (2018): Quantifying uncertainty from aerosol and atmospheric parameters and their impact on climate sensitivity. *Atmos. Chem. Phys.* **18**(23), pp. 17529–17543. doi: 10.5194/acp-18-17529-2018. url: https://doi.org/10.5194/acp-18-17529-2018.

[27] Flohn, H. (1964): *Investigations on the Tropical Easterly Jet*. Bonner meteorologische Abhandlungen. Dümmlers.

[28] Forrester, A. I. J., A. Sóbester, and A. J. Keane (2008): *Engineering Design via Surrogate Modelling: A Practical Guide*. University of Southampton, UK: Wiley. doi: 10.1002/9780470770801.

[29] Forrester, A. I. and A. J. Keane (2009): Recent advances in surrogate-based optimization. *Progress in Aerospace Sciences* **45**(1), pp. 50–79. doi: 10.1016/j.paerosci.2008.11.001.

[30] Fotso-Nguemo, T. C., D. A. Vondou, W. M. Pokam, Z. Y. Djomou, I. Diallo, A. Haensler, L. A. D. Tchotchou, P. H. Kamsu-Tamo, A. T. Gaye, and C. Tchawoua (2017): On the added value of the regional climate model REMO in the assessment of climate change signal over Central Africa. *Climate Dynamics* **49**, pp. 3813–3838. doi: 10.1007/s00382-017-3547-7.

[31] Garud, S. S., I. A. Karimi, and M. Kraft (2017): Design of computer experiments: A review. *Computers & Chemical Engineering* **106**, pp. 71–95. doi: 10.1016/j.compchemeng.2017.05.010.

[32] Gbode, I. E., J. Dudhia, K. O. Ogunjobi, and V. O. Ajayi (2018): Sensitivity of different physics schemes in the WRF model during a West African monsoon regime. *Theoretical and Applied Climatology* **136**(1-2), pp. 733–751. doi: 10.1007/s00704-018-2538-x.

[33] Gbode, I. E., T. E. Babalola, G. T. Diro, and J. D. Intsiful (2023): Assessment of ERA5 and ERA-Interim in Reproducing Mean and Extreme Climates over West Africa. *Adv. Atmos. Sci.* **40**(4), pp. 570–586. doi: 10.1007/s00376-022-2161-8.

[34] Geladi, P. and B. R. Kowalski (1986): Partial least-squares regression: a tutorial. *Analytica Chimica Acta* **185**, pp. 1–17. doi: 10.1016/0003-2670(86)80028-9.

[35] Geladi, P., H. Isaksson, L. Lindqvist, S. Wold, and K. Esbensen (1989): Principal component analysis of multivariate images. *Chemometrics and Intelligent Laboratory Systems* **5**(3), pp. 209–220. doi: 10.1016/0169-7439(89)80049-8.

[36] Geladi, P. and K. Esbensen (1991): Regression on multivariate images: Principal component regression for modeling, prediction and visual diagnostic tools. *Journal of Chemometrics* **5**(2), pp. 97–111. doi: 10.1002/cem.1180050206.

[37] Gibbs, M. N. (1998): *Bayesian Gaussian processes for regression and classification.* PhD thesis. University of Cambridge.

[38] Gill, A. E. (1980): Some simple solutions for heat-induced tropical circulation. *Quarterly Journal of the Royal Meteorological Society* **106**(449), pp. 447–462. doi: 10.1002/qj.49710644905.

[39] Glassmeier, F., F. Hoffmann, J. S. Johnson, T. Yamaguchi, K. S. Carslaw, and G. Feingold (2019): An emulator approach to stratocumulus susceptibility. *Atmos. Chem. Phys.* **19**(15), pp. 10191–10203. doi: 10.5194/acp-19-10191-2019. url: https://doi.org/10.5194/acp-19-10191-2019.

[40] Golding, B., K. Mylne, and P. Clark (2004): The history and future of numerical weather prediction in the Met Office. *Weather* **59**(11), pp. 299–306. doi: 10.1256/wea.113.04.

[41] Gong, W., Q. Duan, J. Li, C. Wang, Z. Di, A. Ye, C. Miao, and Y. Dai (2016): Multiobjective adaptive surrogate modeling-based optimization for parameter estimation of large, complex geophysical models. *Water Resources Research* **52**(3), pp. 1984–2008. doi: 10.1002/2015wr018230.

[42] Grist, J. P. and S. E. Nicholson (2001): A Study of the Dynamic Factors Influencing the Rainfall Variability in the West African Sahel. *Journal of Climate* **14**(7), pp. 1337–1359. doi: 10.1175/1520-0442(2001)014<1337:asotdf>2.0.co;2.

[43] Haile, M. (2005): Weather patterns, food security and humanitarian response in sub-Saharan Africa. *Philosophical Transactions of the Royal Society B: Biological Sciences* **360**(1463), pp. 2169–2182. doi: 10.1098/rstb.2005.1746.

[44] Hall, N. M. and P. Peyrillé (2006): Dynamics of the West African monsoon. *Journal de Physique IV (Proceedings)* **139**(1), pp. 81–99. doi: 10.1051/jp4:2006139007.

[45] Hannak, L., P. Knippertz, A. H. Fink, A. Kniffka, and G. Pante (2017): Why Do Global Climate Models Struggle to Represent Low-Level Clouds in the West African Summer Monsoon? *Journal of Climate* **30**(5), pp. 1665–1687. doi: 10.1175/jcli-d-16-0451.1.

[46] Harman, H. H. (1976): *Modern Factor Analysis*. The University of Chicago Press, Chicago.

[47] Harper, W. V. and S. K. Gupta (1983): *Sensitivity/uncertainty analysis of a borehole scenario comparing Latin Hypercube Sampling and deterministic sensitivity approaches*. Tech. rep. Battelle Memorial Inst., Columbus, OH (USA). Office of Nuclear Waste Isolation.

[48] Hastenrath, S. (1991): *Climate Dynamics of the Tropics*. Springer Netherlands. doi: 10.1007/978-94-011-3156-8.

[49] Hersbach, H., B. Bell, P. Berrisford, S. Hirahara, A. Horányi, J. Muñoz-Sabater, J. Nicolas, C. Peubey, R. Radu, D. Schepers, A. Simmons, C. Soci, S. Abdalla, X. Abellan, G. Balsamo, P. Bechtold, G. Biavati, J. Bidlot, M. Bonavita, G. Chiara, P. Dahlgren, D. Dee, M. Diamantakis, R. Dragani, J. Flemming, R. Forbes, M. Fuentes, A. Geer, L. Haimberger, S. Healy, R. J. Hogan, E. Hólm, M. Janisková, S. Keeley, P. Laloyaux, P. Lopez, C. Lupu, G. Radnoti, P. Rosnay, I. Rozum, F. Vamborg, S. Villaume, and J.-N. Thépaut (2020): The ERA5 global reanalysis. *Quarterly Journal of the Royal Meteorological Society* **146**(730), pp. 1999–2049. doi: 10.1002/qj.3803.

[50] Holden, P. B., N. R. Edwards, K. I. C. Oliver, T. M. Lenton, and R. D. Wilkinson (2009): A probabilistic calibration of climate sensitivity and terrestrial carbon change in GENIE-1. *Climate Dynamics* **35**(5), pp. 785–806. doi: 10.1007/s00382-009-0630-8. url: https://doi.org/10.1007/s00382-009-0630-8.

[51] Hopcroft, P. O., P. J. Valdes, A. B. Harper, and D. J. Beerling (2017): Multi vegetation model evaluation of the Green Sahara climate regime. *Geophysical Research Letters* **44**(13), pp. 6804–6813. doi: 10.1002/2017gl073740.

[52] Huffman, G., E. Stocker, D. Bolvin, E. Nelkin, and J. Tan (2019): GPM IMERG final precipitation L3 half hourly 0.1 degree x 0.1 degree V06. *Goddard Earth Sciences Data and Information Services Center (GES DISC): Greenbelt, MD, USA.*

[53] Hunt, B. R., E. J. Kostelich, and I. Szunyogh (2007): Efficient data assimilation for spatiotemporal chaos: A local ensemble transform Kalman filter. *Physica D: Nonlinear Phenomena* **230**(1), pp. 112–126. doi: 10.1016/j.physd.2006.11.008.

[54] Janicot, S., J.-P. Lafore, and C. Thorncroft (2011): The West African Monsoon. In: *The Global Monsoon System.* World Scientific, pp. 111–135. doi: 10.1142/9789814343411_0008.

[55] John D. Kelleher, B. M. N. and A. D'Arcy (2015): *Fundamentals of Machine Learning for Predictive Data Analytics.* The MIT Press.

[56] Johnson, M., L. Moore, and D. Ylvisaker (1990): Minimax and maximin distance designs. *Journal of Statistical Planning and Inference* **26**(2), pp. 131–148. doi: 10.1016/0378-3758(90)90122-B.

[57] Jolliffe, I. T. (1986): *Principal Component Analysis.* Springer New York. doi: 10.1007/978-1-4757-1904-8.

[58] Joseph, V. R., Y. Hung, and A. Sudjianto (2008): Blind Kriging: A New Method for Developing Metamodels. *Journal of Mechanical Design* **130**(3), p. 031102. doi: 10.1115/1.2829873.

[59] Kendon, E. J., R. A. Stratton, S. Tucker, J. H. Marsham, S. Berthou, D. P. Rowell, and C. A. Senior (2019): Enhanced future changes in wet and dry extremes over Africa at convection-permitting scale. *Nature Communications* **10**(1). doi: 10.1038/s41467-019-09776-9.

[60] Kersaudy, P., B. Sudret, N. Varsier, O. Picon, and J. Wiart (2015): A new surrogate modeling technique combining Kriging and polynomial chaos expansions – Application to uncertainty analysis in computational dosimetry. *Journal of Computational Physics* **286**, pp. 103–117. doi: 10.1016/j.jcp.2015.01.034.

[61] Kiladis, G. N., C. D. Thorncroft, and N. M. J. Hall (2006): Three-Dimensional Structure and Dynamics of African Easterly Waves. Part I: Observations. *Journal of the Atmospheric Sciences* **63**(9), pp. 2212–2230. doi: 10.1175/jas3741.1.

[62] Klein, C., D. Heinzeller, J. Bliefernicht, and H. Kunstmann (2015): Variability of West African monsoon patterns generated by a WRF multi-physics ensemble. *Climate Dynamics* **45**(9-10), pp. 2733–2755. doi: 10.1007/s00382-015-2505-5.

[63] Kniffka, A., P. Knippertz, and A. H. Fink (2019): The role of low-level clouds in the West African monsoon system. *Atmos. Chem. Phys.* **19**(3), pp. 1623–1647. doi: 10.5194/acp-19-1623-2019.

[64] Kniffka, A., P. Knippertz, A. H. Fink, A. Benedetti, M. E. Brooks, P. G. Hill, M. Maranan, G. Pante, and B. Vogel (2020): An evaluation of operational and research weather forecasts for southern West Africa using observations from

the DACCIWA field campaign in June–July 2016. *Quarterly Journal of the Royal Meteorological Society* **146**(728), pp. 1121–1148. DOI: 10.1002/qj.3729.

[65] KNIPPERTZ, P., A. ANSMANN, D. ALTHAUSEN, D. MÜLLER, M. TESCHE, E. BIERWIRTH, T. DINTER, T. MÜLLER, W. V. HOYNINGEN-HUENE, K. SCHEPANSKI, M. WENDISCH, B. HEINOLD, K. KANDLER, A. PETZOLD, L. SCHÜTZ, and I. TEGEN (2009): Dust mobilization and transport in the northern Sahara during SAMUM 2006 – a meteorological overview. *Tellus B: Chemical and Physical Meteorology* **61**(1), p. 12. DOI: 10.1111/j.1600-0889.2008.00380.x.

[66] LANG, S. T. K., S.-J. LOCK, M. LEUTBECHER, P. BECHTOLD, and R. M. FORBES (2021): Revision of the Stochastically Perturbed Parametrisations model uncertainty scheme in the Integrated Forecasting System. *Quarterly Journal of the Royal Meteorological Society* **147**(735), pp. 1364–1381. DOI: 10.1002/qj.3978.

[67] LATANIOTIS, C., S. MARELLI, and B. SUDRET (2018): The Gaussian Process Modeling Module in UQLab. *Journal of Soft Computing in Civil Engineering* **2**(3), pp. 91–116. DOI: 10.22115/scce.2018.129323.1062.

[68] LAVAYSSE, C., C. FLAMANT, S. JANICOT, D. J. PARKER, J.-P. LAFORE, B. SULTAN, and J. PELON (2009): Seasonal evolution of the West African heat low: a climatological perspective. *Climate Dynamics* **33**(2-3), pp. 313–330. DOI: 10.1007/s00382-009-0553-4.

[69] LEBEL, T., A. DIEDHIOU, and H. LAURENT (2003): Seasonal cycle and interannual variability of the Sahelian rainfall at hydrological scales. *Journal of Geophysical Research* **108**(D8). DOI: 10.1029/2001jd001580.

[70] LEBEL, T. and A. ALI (2009): Recent trends in the Central and Western Sahel rainfall regime (1990–2007). *Journal of Hydrology* **375**(1-2), pp. 52–64. DOI: 10.1016/j.jhydrol.2008.11.030.

[71] LEE, L. A., K. S. CARSLAW, K. J. PRINGLE, G. W. MANN, and D. V. SPRACKLEN (2011): Emulation of a complex global aerosol model to quantify sensitivity to uncertain parameters. *Atmos. Chem. Phys.* **11**, pp. 12253–12273. DOI: 10.5194/acp-11-12253-2011.

[72] LEE, L. A., C. L. REDDINGTON, and K. S. CARSLAW (2016): On the relationship between aerosol model uncertainty and radiative forcing uncertainty. *Proceedings of the National Academy of Sciences* **113**(21), pp. 5820–5827. DOI: 10.1073/pnas.1507050113.

[73] LEMAIRE, M. (2009): *Structural Reliability*. John Wiley & Sons.

[74] LEMBURG, A., J. BADER, and M. CLAUSSEN (2019): Sahel Rainfall–Tropical Easterly Jet Relationship on Synoptic to Intraseasonal Time Scales. *Monthly Weather Review* **147**(5), pp. 1733–1752. DOI: 10.1175/mwr-d-18-0254.1.

[75] VAN DER LINDEN, R., P. KNIPPERTZ, A. H. FINK, B. INGLEBY, M. MARANAN, and A. BENEDETTI (2020): The influence of DACCIWA radiosonde data on the quality

of ECMWF analyses and forecasts over southern West Africa. *Quarterly Journal of the Royal Meteorological Society* **146**(729), pp. 1719–1739. DOI: 10.1002/qj.3763.

[76] LOEPPKY, J. L., J. SACKS, and W. J. WELCH (2009): Choosing the Sample Size of a Computer Experiment: A Practical Guide. *Technometrics* **51**(4), pp. 366–376. DOI: 10.1198/tech.2009.08040.

[77] LOHOU, F., N. KALTHOFF, B. ADLER, K. BABIĆ, C. DIONE, M. LOTHON, X. PEDRUZO-BAGAZGOITIA, and M. ZOUZOUA (2020): Conceptual model of diurnal cycle of low-level stratiform clouds over southern West Africa. *Atmos. Chem. Phys.* **20**(4), pp. 2263–2275. DOI: 10.5194/acp-20-2263-2020.

[78] LU, D., D. RICCIUTO, M. STOYANOV, and L. GU (2018): Calibration of the E3SM Land Model Using Surrogate-Based Global Optimization. *Journal of Advances in Modeling Earth Systems* **10**, pp. 1337–1356. DOI: 10.1002/2017MS001134.

[79] LU, D. and D. RICCIUTO (2019): Efficient surrogate modeling methods for large-scale Earth system models based on machine-learning techniques. *Geoscientific Model Development* **12**(5), pp. 1791–1807. DOI: 10.5194/gmd-12-1791-2019.

[80] LU, L., C. M. ANDERSON-COOK, and T. AHMED (2021): Non-uniform space filling (NUSF) designs. *Journal of Quality Technology* **53**(3), pp. 309–330. DOI: 10.1080/00224065.2020.1727801.

[81] M. D. MCKAY, R. J. B. and W. J. CONOVER (1979): Comparison of Three Methods for Selecting Values of Input Variables in the Analysis of Output from a Computer Code. *Technometrics* **21**(2), pp. 239–245. DOI: 10.1080/00401706.1979.10489755.

[82] MARSHALL, A. W. (1954): *The Use of Multistage Sampling Schemes in Monte Carlo Computations*. Santa Monica, California: Rand Corporation.

[83] MARSHAM, J. H., N. S. DIXON, L. GARCIA-CARRERAS, G. M. S. LISTER, D. J. PARKER, P. KNIPPERTZ, and C. E. BIRCH (2013): The role of moist convection in the West African monsoon system: Insights from continental-scale convection-permitting simulations. *Geophysical Research Letters* **40**(9), pp. 1843–1849. DOI: 10.1002/grl.50347.

[84] MARTIN, G. M., P. PEYRILLÉ, R. ROEHRIG, C. RIO, M. CAIAN, G. BELLON, F. CODRON, J.-P. LAFORE, D. E. POAN, and A. IDELKADI (2017): Understanding the West African Monsoon from the analysis of diabatic heating distributions as simulated by climate models. *Journal of Advances in Modeling Earth Systems* **9**(1), pp. 239–270. DOI: 10.1002/2016ms000697.

[85] MASSOUD, E. C. (2019): Emulation of environmental models using polynomial chaos expansion. *Environmental Modelling & Software* **111**, pp. 421–431. DOI: 10.1016/j.envsoft.2018.10.008.

[86] MATHERON, G. (1962): *Traité de géostatistique appliquée, tome I*. Mémoires du Bureau de Recherches Géologiques et Minières 14. Paris: Éditions Technip.

[87] Matheron, G. (1969): *Le krigeage universel*. Vol. 1. Fontainebleau: Cahiers du Centre de Morphologie Mathématique, École des Mines de Paris, NA.

[88] Matheron, G. (1971): *The Theory of Regionalized Variables and Its Applications*. École Nationale Supérieure des Mines de Paris.

[89] Mathon, V., H. Laurent, and T. Lebel (2002): Mesoscale Convective System Rainfall in the Sahel. *Journal of Applied Meteorology* **41**(11), pp. 1081–1092. doi: 10.1175/1520-0450(2002)041<1081:mcsrit>2.0.co;2.

[90] Matsui, T., S. Q. Zhang, S. E. Lang, W.-K. Tao, C. Ichoku, and C. D. Peters-Lidard (2018): Impact of radiation frequency, precipitation radiative forcing, and radiation column aggregation on convection-permitting West African monsoon simulations. *Climate Dynamics* **55**(1-2), pp. 193–213. doi: 10.1007/s00382-018-4187-2.

[91] Melchers, R. E. and A. T. Beck (2017): Rosenblatt and Other Transformations. In: *Structural Reliability Analysis and Prediction*. John Wiley & Sons, Ltd. Chap. B, pp. 403–414.

[92] Messager, C., H. Gallée, and O. Brasseur (2004): Precipitation sensitivity to regional SST in a regional climate simulation during the West African monsoon for two dry years. *Climate Dynamics* **22**(2-3), pp. 249–266. doi: 10.1007/s00382-003-0381-x.

[93] Moore, G. E. (1965): Cramming More Components onto Integrated Circuits. *Electronics Magazine* **38**(8), pp. 114–117. url: http://cva.stanford.edu/classes/cs99s/papers/moore-crammingmorecomponents.pdf (visited on 03/18/2025).

[94] Morris, M. D., T. J. Mitchell, and D. Ylvisaker (1993): Bayesian Design and Analysis of Computer Experiments: Use of Derivatives in Surface Prediction. *Technometrics* **35**(3), pp. 243–255. doi: 10.1080/00401706.1993.10485320.

[95] Morris, M. D. and T. J. Mitchell (1995): Exploratory designs for computational experiments. *Journal of Statistical Planning and Inference* **43**(3), pp. 381–402. doi: 10.1016/0378-3758(94)00035-T.

[96] Moustapha, M., B. Sudret, J.-M. Bourinet, and B. Guillaume (2016): Quantile-based optimization under uncertainties using adaptive Kriging surrogate models. *Structural and Multidisciplinary Optimization* **54**(6), pp. 1403–1421. doi: 10.1007/s00158-016-1504-4.

[97] Müller, J., R. Paudel, C. A. Shoemaker, J. Woodbury, Y. Wang, and N. Mahowald (2015): $CH_4$ parameter estimation in CLM4.5bgc using surrogate global optimization. *Geoscientific Model Development* **8**, pp. 3285–3310. doi: 10.5194/gmd-8-3285-2015.

[98] Næs, T. and H. Martens (1988): Principal component regression in NIR analysis: Viewpoints, background details and selection of components. *Journal of Chemometrics* **2**(2), pp. 155–167. doi: 10.1002/cem.1180020207.

[99] Neelin, J. D., A. Bracco, H. Luo, J. C. McWilliams, and J. E. Meyerson (2010): Considerations for parameter optimization and sensitivity in climate models. *Proceedings of the National Academy of Sciences* **107**(50), pp. 21349–21354. doi: 10.1073/pnas.1015473107.

[100] Nicholson, S. E. (2009): A revised picture of the structure of the "monsoon" and land ITCZ over West Africa. *Climate Dynamics* **32**, pp. 1155–1171. doi: 10.1007/s00382-008-0514-3.

[101] Oakley, J. and A. O'Hagan (2002): Bayesian inference for the uncertainty distribution of computer model outputs. *Biometrika* **89**(4), pp. 769–784. doi: 10.1093/biomet/89.4.769.

[102] Oakley, J. (2004): Estimating percentiles of uncertain computer code outputs. *Journal of the Royal Statistical Society: Series C (Applied Statistics)* **53**(1), pp. 83–93. doi: 10.1046/j.0035-9254.2003.05044.x.

[103] Oakley, J. E. and A. O'Hagan (2004): Probabilistic sensitivity analysis of complex models: a Bayesian approach. *Journal of the Royal Statistical Society: Series B (Statistical Methodology)* **66**(3), pp. 751–769. doi: 10.1111/j.1467-9868.2004.05304.x.

[104] Ollinaho, P., H. Järvinen, P. Bauer, M. Laine, P. Bechtold, J. Susiluoto, and H. Haario (2014): Optimization of NWP model closure parameters using total energy norm of forecast error as a target. *Geoscientific Model Development* **7**(5), pp. 1889–1900. doi: 10.5194/gmd-7-1889-2014.

[105] Ollinaho, P., S.-J. Lock, M. Leutbecher, P. Bechtold, A. Beljaars, A. Bozzo, R. M. Forbes, T. Haiden, R. J. Hogan, and I. Sandu (2017): Towards process-level representation of model uncertainties: stochastically perturbed parametrizations in the ECMWF ensemble. *Quarterly Journal of the Royal Meteorological Society* **143**(702), pp. 408–422. doi: 10.1002/qj.2931.

[106] Paciorek, C. and M. Schervish (2003): Nonstationary Covariance Functions for Gaussian Process Regression. In: *Advances in Neural Information Processing Systems*. S. Thrun, L. Saul, and B. Schölkopf eds. Vol. 16. MIT Press.

[107] Paeth, H., A. Capo-Chichi, and W. Endlicher (2008): Climate change and food security in tropical West Africa – a dynamic-statistical modelling approach. *Erdkunde* **62**, pp. 101–115. doi: 10.3112/erdkunde.2008.02.01.

[108] Pante, G. and P. Knippertz (2019): Resolving Sahelian thunderstorms improves mid-latitude weather forecasts. *Nature Communications* **10**(1). doi: 10.1038/s41467-019-11081-4.

[109] Parker, D. J., A. Fink, S. Janicot, J.-B. Ngamini, M. Douglas, E. Afiesimama, A. Agusti-Panareda, A. Beljaars, F. Dide, A. Diedhiou, T. Lebel, J. Polcher, J.-L. Redelsperger, C. Thorncroft, and G. A. Wilson (2008): The AMMA Radiosonde Program and its Implications for the Future of Atmospheric Monitoring Over Africa. *Bulletin of the American Meteorological Society* **89**(7), pp. 1015–1028. doi: 10.1175/2008bams2436.1.

[110] Pronzato, L. and W. G. Müller (2011): Design of computer experiments: space filling and beyond. *Statistics and Computing* **22**(3), pp. 681–701. doi: 10.1007/s11222-011-9242-3. url: http://dx.doi.org/10.1007/s11222-011-9242-3.

[111] Quinlan, K. R., J. Movva, and B. Perfect (2024): Non-uniform active learning for Gaussian process models with applications to trajectory informed aerodynamic databases. *Statistical Analysis and Data Mining: The ASA Data Science Journal* **17**(2), e11675. doi: 10.1002/sam.11675.

[112] Raschendorfer, M. (2012): *Operationelles NWV-System, hier: Verminderung der minimalen Diffusionskoeffizienten für COSMO-EU/DE/EPS*. Tech. rep. url: https://www.dwd.de/DE/fachnutzer/forschung_lehre/numerische_wettervorhersage/nwv_aenderungen/_functions/DownloadBox_modellaenderungen/cosmo_de/pdf_2011_2015/pdf_cosmo_de_29_11_2012.pdf?__blob=publicationFile&v=4 (visited on 12/18/2023).

[113] Rasmussen, C. E. and C. K. I. Williams (2005): *Gaussian Processes for Machine Learning*. The MIT Press. doi: 10.7551/mitpress/3206.001.0001.

[114] Ray, J., Z. Hou, M. Huang, K. Sargsyan, and L. Swiler (2015): Bayesian Calibration of the Community Land Model Using Surrogates. *SIAM/ASA Journal on Uncertainty Quantification* **3**, pp. 199–233. doi: 10.1137/140957998.

[115] Reed, R. J., D. C. Norquist, and E. E. Recker (1977): The Structure and Properties of African Wave Disturbances as Observed During Phase III of GATE. *Monthly Weather Review* **105**(3), pp. 317–333. doi: 10.1175/1520-0493(1977)105<0317:tsapoa>2.0.co;2.

[116] Reinert, D., F. Prill, D. M., and G. Zängl (2019): *Database Reference Manual for ICON and ICON-EPS, Version 1.2.11*. Tech. rep.

[117] Rosenblatt, M. (1952): Remarks on a Multivariate Transformation. *The Annals of Mathematical Statistics* **23**(3), pp. 470–472. doi: 10.1214/aoms/1177729394.

[118] Ruiz, J. J., M. Pulido, and T. Miyoshi (2013): Estimating Model Parameters with Ensemble-Based Data Assimilation: A Review. *Journal of the Meteorological Society of Japan. Ser. II* **91**(2), pp. 79–99. doi: 10.2151/jmsj.2013-201.

[119] Sacks, J., W. J. Welch, T. J. Mitchell, and H. P. Wynn (1989): Design and Analysis of Computer Experiments. *Statistical Science* **4**(4), pp. 409–423. doi: 10.1214/ss/1177012413.

[120] Saltelli, A., K. Chan, and E. M. Scott (2000): *Sensitivity Analysis*. John Wiley & Sons.

[121] Sampson, P. D. and P. Guttorp (1992): Nonparametric estimation of nonstationary spatial covariance structure. *Journal of the American Statistical Association* **87**(417), pp. 108–119. doi: 10.2307/2290458.

[122] Santner, T. J., B. J. Williams, and W. I. Notz (2003): *The Design and Analysis of Computer Experiments*. Springer New York. doi: 10.1007/978-1-4757-3799-8.

[123] Schmidt, A. M. and A. O'Hagan (2003): Bayesian inference for non-stationary spatial covariance structure via spatial deformations. *Journal of the Royal Statistical Society: Series B (Statistical Methodology)* **65**(3), pp. 743–758. doi: 10.1111/1467-9868.00413.

[124] Shafii, M. and F. De Smedt (2009): Multi-objective calibration of a distributed hydrological model (WetSpa) using a genetic algorithm. *Hydrology and Earth System Sciences* **13**(11), pp. 2137–2149. doi: 10.5194/hess-13-2137-2009.

[125] Sobol, I. M. (1993): Sensitivity Estimates for Nonlinear Mathematical Models. *Mathematical Modelling and Computational Experiments* **1**, pp. 407–414.

[126] Stubenrauch, C. J., G. Caria, S. E. Protopapadaki, and F. Hemmer (2021): 3D radiative heating of tropical upper tropospheric cloud systems derived from synergistic A-Train observations and machine learning. *Atmos. Chem. Phys.* **21**(2), pp. 1015–1034. doi: 10.5194/acp-21-1015-2021.

[127] Sudret, B. (2014): Polynomial chaos expansions and stochastic finite element methods. In: *Risk and Reliability in Geotechnical Engineering*. K.-K. Phoon and J. Ching eds. London: CRC Press, pp. 265–300.

[128] Sudret, B., S. Marelli, and J. Wiart (2017): Surrogate models for uncertainty quantification: An overview. In: *2017 11th European Conference on Antennas and Propagation (EUCAP)*. IEEE, pp. 793–797. doi: 10.23919/eucap.2017.7928679.

[129] Sullivan, T. (2015): *Introduction to Uncertainty Quantification*. Vol. 63. Springer International Publishing. doi: 10.1007/978-3-319-23395-6.

[130] Tatang, M., W. Pan, R. Prinn, and G. McRae (1997): An Efficient Method for Parametric Uncertainty Analysis of Numerical Geophysical Model. *Journal of Geophysical Research* **102**, pp. 21925–21932. doi: 10.1029/97JD01654.

[131] Tchotchou, L. A. D. and F. M. Kamga (2009): Sensitivity of the simulated African monsoon of summers 1993 and 1999 to convective parameterization schemes in RegCM3. *Theoretical and Applied Climatology* **100**(1-2), pp. 207–220. doi: 10.1007/s00704-009-0181-2.

[132] Thorncroft, C. D., H. Nguyen, C. Zhang, and P. Peyrillé (2011): Annual cycle of the West African monsoon: regional circulations and associated water vapour transport. *Quarterly Journal of the Royal Meteorological Society* **137**(654), pp. 129–147. doi: 10.1002/qj.728.

[133] V. Roshan Joseph Li Gu, S. B. and W. R. Myers (2019): Space-Filling Designs for Robustness Experiments. *Technometrics* **61**(1), pp. 24–37. doi: 10.1080/00401706.2018.1451390.

[134] Vapnik, V. (1995): *The Nature of Statistical Learning Theory*. Springer New York. doi: 10.1007/978-1-4757-2440-0.

[135] Vellinga, M., A. Arribas, and R. Graham (2013): Seasonal forecasts for regional onset of the West African monsoon. *Climate Dynamics* **40**, pp. 3047–3070. doi: 10.1007/s00382-012-1520-z.

[136] Viana, F. A. C., C. Gogu, and T. Goel (2021): Surrogate modeling: tricks that endured the test of time and some recent developments. *Structural and Multidisciplinary Optimization* **64**, pp. 2881–2908. doi: 10.1007/s00158-021-03001-2.

[137] Virtanen, P., R. Gommers, T. E. Oliphant, M. Haberland, T. Reddy, D. Cournapeau, E. Burovski, P. Peterson, W. Weckesser, J. Bright, S. J. van der Walt, M. Brett, J. Wilson, K. J. Millman, N. Mayorov, A. R. J. Nelson, E. Jones, R. Kern, E. Larson, C. J. Carey, İ. Polat, Y. Feng, E. W. Moore, J. VanderPlas, D. Laxalde, J. Perktold, R. Cimrman, I. Henriksen, E. A. Quintero, C. R. Harris, A. M. Archibald, A. H. Ribeiro, F. Pedregosa, P. van Mulbregt, and SciPy 1.0 Contributors (2020): SciPy 1.0: Fundamental Algorithms for Scientific Computing in Python. *Nature Methods* **17**, pp. 261–272. doi: 10.1038/s41592-019-0686-2.

[138] Vogel, P., P. Knippertz, A. H. Fink, A. Schlueter, and T. Gneiting (2018): Skill of Global Raw and Postprocessed Ensemble Predictions of Rainfall over Northern Tropical Africa. *Weather and Forecasting* **33**(2), pp. 369–388. doi: 10.1175/waf-d-17-0127.1.

[139] Vogel, P., P. Knippertz, A. H. Fink, A. Schlueter, and T. Gneiting (2020): Skill of Global Raw and Postprocessed Ensemble Predictions of Rainfall in the Tropics. *Weather and Forecasting* **35**(6), pp. 2367–2385. doi: 10.1175/waf-d-20-0082.1.

[140] Walz, E., M. Maranan, R. van der Linden, A. H. Fink, and P. Knippertz (2021): An IMERG-Based Optimal Extended Probabilistic Climatology (EPC) as a Benchmark Ensemble Forecast for Precipitation in the Tropics and Subtropics. *Weather and Forecasting*. doi: 10.1175/waf-d-20-0233.1.

[141] Wan, H., P. J. Rasch, K. Zhang, Y. Qian, H. Yan, and C. Zhao (2014): Short ensembles: an efficient method for discerning climate-relevant sensitivities in atmospheric general circulation models. *Geoscientific Model Development* **7**(5), pp. 1961–1977. doi: 10.5194/gmd-7-1961-2014.

[142] Watson-Parris, D., A. Williams, L. Deaconu, and P. Stier (2021): Model calibration using ESEm v1.1.0 – an open, scalable Earth system emulator. *Geoscientific Model Development* **14**, pp. 7659–7672. doi: 10.5194/gmd-14-7659-2021.

[143] Webster, M. D., M. A. Tatang, and G. J. McRae (1996): Application of Probabilistic Collocation Method for Uncertainty Analysis of a Simple Ocean Model.

*Joint Program Report Series, Report 4.* URL: http://globalchange.mit.edu/publication/15670 (visited on 01/11/2023).

[144] Wellmann, C., A. I. Barrett, J. S. Johnson, M. Kunz, B. Vogel, K. S. Carslaw, and C. Hoose (2020): Comparing the impact of environmental conditions and microphysics on the forecast uncertainty of deep convective clouds and hail. *Atmos. Chem. Phys.* **20**(4), pp. 2201–2219. DOI: 10.5194/acp-20-2201-2020.

[145] Williamson, D., M. Goldstein, L. Allison, A. Blaker, P. Challenor, L. Jackson, and K. Yamazaki (2013): History matching for exploring and reducing climate model parameter space using observations and a large perturbed physics ensemble. *Climate Dynamics* **41**, pp. 1703–1729. DOI: 10.1007/s00382-013-1896-4.

[146] Williamson, D. (2015): Exploratory ensemble designs for environmental models using k-extended Latin Hypercubes. *Environmetrics* **26**, pp. 268–283. DOI: 10.1002/env.2335.

[147] Xiong, Y., W. Chen, D. Apley, and X. Ding (2007): A non-stationary covariance-based Kriging method for metamodelling in engineering design. *International Journal for Numerical Methods in Engineering* **71**(6), pp. 733–756. DOI: 10.1002/nme.1969.

[148] Xiu, D. and G. E. Karniadakis (2002): The Wiener-Askey Polynomial Chaos for Stochastic Differential Equations. *SIAM Journal on Scientific Computing* **24**(2), pp. 619–644. DOI: 10.1137/s1064827501387826.

[149] Xu, D., Y. Shi, I. W. Tsang, Y.-S. Ong, C. Gong, and X. Shen (2020): Survey on Multi-Output Learning. *IEEE Transactions on Neural Networks and Learning Systems* **31**(7), pp. 2409–2429. DOI: 10.1109/TNNLS.2019.2945133.

[150] Xu, S., X. An, X. Qiao, L. Zhu, and L. Li (2013): Multi-output least-squares support vector regression machines. *Pattern Recognition Letters* **34**(9), pp. 1078–1084. DOI: 10.1016/j.patrec.2013.01.015.

[151] Xue, Y., F. D. Sales, W. K.-M. Lau, A. Boone, J. Feng, P. Dirmeyer, Z. Guo, K.-M. Kim, A. Kitoh, V. Kumar, I. Poccard-Leclercq, N. Mahowald, W. Moufouma-Okia, P. Pegion, D. P. Rowell, J. Schemm, S. D. Schubert, A. Sealy, W. M. Thiaw, A. Vintzileos, S. F. Williams, and M.-L. C. Wu (2010): Intercomparison and analyses of the climatology of the West African Monsoon in the West African Monsoon Modeling and Evaluation project (WAMME) first model intercomparison experiment. *Climate Dynamics* **35**, pp. 3–27. DOI: 10.1007/s00382-010-0778-2.

[152] Zängl, G., D. Reinert, P. Rípodas, and M. Baldauf (2015): The ICON (ICOsahedral Non-hydrostatic) modelling framework of DWD and MPI-M: Description of the non-hydrostatic dynamical core. *Quarterly Journal of the Royal Meteorological Society* **141**(687), pp. 563–579. DOI: 10.1002/qj.2378.

[153] Zängl, G. (2023): Adaptive tuning of uncertain parameters in a numerical weather prediction model based upon data assimilation. *Quarterly Journal of the Royal Meteorological Society* **149**(756), pp. 2861–2880. doi: 10.1002/qj.4535.

[154] Zhang, J., A. Messac, J. Zhang, and S. Chowdhury (2012): Improving the Accuracy of Surrogate Models Using Inverse Transform Sampling. In: *53rd AIAA/ASME/ASCE/AHS/ASC Structures, Structural Dynamics and Materials Conference 20th AIAA/ASME/AHS Adaptive Structures Conference 14th AIAA*. American Institute of Aeronautics and Astronautics, p. 1429. doi: 10.2514/6.2012-1429.

[155] Zheng, X. and E. A. B. Eltahir (1998): The Role of Vegetation in the Dynamics of West African Monsoons. *Journal of Climate* **11**, pp. 2078–2096.

[156] Zou, H. and T. Hastie (2005): Regularization and Variable Selection Via the Elastic Net. *Journal of the Royal Statistical Society Series B: Statistical Methodology* **67**(2), pp. 301–320. doi: 10.1111/j.1467-9868.2005.00503.x.

# Publications


[157] Fischer, M. and C. Proppe (2021): A sequential multi-point sampling procedure for surrogate models. In: *4th International Conference on Uncertainty Quantification in Computational Sciences and Engineering*. Institute of Research and Development for Computational Methods in Engineering Sciences (ICMES), pp. 246–257. doi: 10.7712/120221.8036.18925.

[158] Fischer, M. and C. Proppe (2023): Enhanced universal kriging for transformed input parameter spaces. *Probabilistic Engineering Mechanics* **74**, p. 103486. doi: 10.1016/j.probengmech.2023.103486.

[159] Fischer, M. (2024): *ICON parameter influence on output variables: mattfis/wam-simulations: v1.0.2.* doi: 10.5281/ZENODO.11505849. url: https://mattfis.github.io/wam-simulations/.

[160] Fischer, M., P. Knippertz, R. van der Linden, A. Lemburg, G. Pante, C. Proppe, and J. H. Marsham (2024): Quantifying uncertainty in simulations of the West African monsoon with the use of surrogate models. *Weather and Climate Dynamics* **5**(2), pp. 511–536. doi: 10.5194/wcd-5-511-2024. url: https://wcd.copernicus.org/articles/5/511/2024/.

[161] Fischer, M., P. Knippertz, and C. Proppe (2025): Surrogate-based model parameter optimization in simulations of the West African monsoon. *Weather and Climate Dynamics* **6**(1), pp. 113–130. doi: 10.5194/wcd-6-113-2025.

# Supervised Theses

[162] Poisson, M. (2022): *Incorporation of Output Correlations into Multi-Output Support Vector Regression Models to Improve Prediction Accuracy.* (unveröffentlicht). Masterarbeit. Karlsruher Institut für Technologie.

KARLSRUHER INSTITUT FÜR TECHNOLOGIE (KIT)
SCHRIFTENREIHE DES INSTITUTS FÜR TECHNISCHE MECHANIK (ITM)

**ISSN 1614-3914**

**Band 1** **Marcus Simon**
Zur Stabilität dynamischer Systeme mit stochastischer Anregung.
ISBN 3-937300-13-9

**Band 2** **Clemens Reitze**
Closed Loop, Entwicklungsplattform für mechatronische Fahrdynamikregelsysteme.
ISBN 3-937300-19-8

**Band 3** **Martin Georg Cichon**
Zum Einfluß stochastischer Anregungen auf mechanische Systeme.
ISBN 3-86644-003-0

**Band 4** **Rainer Keppler**
Zur Modellierung und Simulation von Mehrkörpersystemen unter Berücksichtigung von Greifkontakt bei Robotern.
ISBN 978-3-86644-092-0

**Band 5** **Bernd Waltersberger**
Strukturdynamik mit ein- und zweiseitigen Bindungen aufgrund reibungsbehafteter Kontakte.
ISBN 978-3-86644-153-8

**Band 6** **Rüdiger Benz**
Fahrzeugsimulation zur Zuverlässigkeitsabsicherung von karosseriefesten Kfz-Komponenten.
ISBN 978-3-86644-197-2

**Band 7** **Pierre Barthels**
Zur Modellierung, dynamischen Simulation und Schwingungsunterdrückung bei nichtglatten, zeitvarianten Balkensystemen.
ISBN 978-3-86644-217-7

Band 8 **Hartmut Hetzler**
Zur Stabilität von Systemen bewegter Kontinua mit Reibkontakten am Beispiel des Bremsenquietschens.
ISBN 978-3-86644-229-0

Band 9 **Frank Dienerowitz**
Der Helixaktor – Zum Konzept eines vorverwundenen Biegeaktors.
ISBN 978-3-86644-232-0

Band 10 **Christian Rudolf**
Piezoelektrische Self-sensing-Aktoren zur Korrektur statischer Verlagerungen.
ISBN 978-3-86644-267-2

Band 11 **Günther Stelzner**
Zur Modellierung und Simulation biomechanischer Mehrkörpersysteme.
ISBN 978-3-86644-340-2

Band 12 **Christian Wetzel**
Zur probabilistischen Betrachtung von Schienen- und Kraftfahrzeugsystemen unter zufälliger Windanregung.
ISBN 978-3-86644-444-7

Band 13 **Wolfgang Stamm**
Modellierung und Simulation von Mehrkörpersystemen mit flächigen Reibkontakten.
ISBN 978-3-86644-605-2

Band 14 **Felix Fritz**
Modellierung von Wälzlagern als generische Maschinenelemente einer Mehrkörpersimulation.
ISBN 978-3-86644-667-0

Band 15 **Aydin Boyaci**
Zum Stabilitäts- und Bifurkationsverhalten hochtouriger Rotoren in Gleitlagern.
ISBN 978-3-86644-780-6

Band 16 **Rugerri Toni Liong**
Application of the cohesive zone model to the analysis of rotors with a transverse crack.
ISBN 978-3-86644-791-2

Band 17 **Ulrich Bittner**
Strukturakustische Optimierung von Axialkolbeneinheiten. Modellbildung, Validierung und Topologieoptimierung.
ISBN 978-3-86644-938-1

Band 18 **Alexander Karmazin**
Time-efficient Simulation of Surface-excited Guided Lamb Wave Propagation in Composites.
ISBN 978-3-86644-935-0

Band 19 **Heike Vogt**
Zum Einfluss von Fahrzeug- und Straßenparametern auf die Ausbildung von Straßenunebenheiten.
ISBN 978-3-7315-0023-0

Band 20 **Laurent Ineichen**
Konzeptvergleich zur Bekämpfung der Torsionsschwingungen im Antriebsstrang eines Kraftfahrzeugs.
ISBN 978-3-7315-0030-8

Band 21 **Sietze van Buuren**
Modeling and simulation of porous journal bearings in multibody systems.
ISBN 978-3-7315-0084-1

Band 22 **Dominik Kern**
Neuartige Drehgelenke für reibungsarme Mechanismen.
ISBN 978-3-7315-0103-9

Band 23 **Nicole Gaus**
Zur Ermittlung eines stochastischen Reibwerts und dessen Einfluss auf reibungserregte Schwingungen.
ISBN 978-3-7315-0118-3

Band 24 **Fabian Bauer**
Optimierung der Energieeffizienz zweibeiniger Roboter durch elastische Kopplungen.
ISBN 978-3-7315-0256-2

---

Band 25 **Benedikt Wiegert**
Nichtlineare Schwingungen von Systemen
mit elastohydrodynamischen Linienkontakten.
ISBN 978-3-7315-0350-7

Band 26 **Arsenty Tikhomolov**
Analytische, numerische und messtechnische
Untersuchung der Dynamik von Fahrzeugkupplungen
am Beispiel des Trennproblems.
ISBN 978-3-7315-0362-0

Band 27 **Daniel Maier**
On the Use of Model Order Reduction Techniques
for the Elastohydrodynamic Contact Problem.
ISBN 978-3-7315-0369-9

Band 28 **Xiaoyu Zhang**
Crosswind stability of vehicles under nonstationary
wind excitation.
ISBN 978-3-7315-0376-7

Band 29 **Jens Deppler**
Ein Beitrag zur viskoelastischen Modellierung nichtholonomer
Bindungsgleichungen.
ISBN 978-3-7315-0548-8

Band 30 **Georg Jehle**
Zur Modellbildung und Simulation reibungserregter
Schwingungen in Pkw-Schaltgetrieben.
ISBN 978-3-7315-0668-3

Band 31 **Joachim Klima**
Lubricant transport towards tribocontact in capillary
surface structures.
ISBN 978-3-7315-0814-4

Band 32 **Gábor Kenderi**
Nonparametric identification of nonlinear dynamic systems.
ISBN 978-3-7315-0834-2

Band 33 **Jan Henrik Schmidt**
An efficient solution procedure for elastohydrodynamic contact problems considering structural dynamics.
ISBN 978-3-7315-0872-4

Band 34 **Ulrich Johannes Römer**
Über den Einfluss der Fußgeometrie auf die Energieeffizienz beim zweibeinigen Gehen.
ISBN 978-3-7315-0887-8

Band 35 **Simon Kapelke**
Zur Beeinflussung reibungsbehafteter Systeme mithilfe überlagerter Schwingungen.
ISBN 978-3-7315-0905-9

Band 36 **Jens Burgert**
On direct and inverse problems related to longitudinal impact of non-uniform elastic rods.
ISBN 978-3-7315-1087-1

Band 37 **Tim Leister**
Dynamics of Rotors on Refrigerant-Lubricated Gas Foil Bearings.
ISBN 978-3-7315-1112-0

Band 38 **Simon Schröders**
Ein Beitrag zur asymptotischen Analyse der hydraulischen Ventilsteuerungen.
ISBN 978-3-7315-1221-9

Band 39 **Lukas Julian Oestringer**
Thermomechanischer Kontakt und stochastische Dynamik gleitender Körper mit rauen Oberflächen.
ISBN 978-3-7315-1297-4

Band 40 **Jimmy Alberto Aramendiz Fuentes**
On the Influence of Piecewise Defined Contact Geometries on Friction Dampers.
ISBN 978-3-7315-1267-7

Band 41 **Matthias A. Fischer**
Surrogate-based uncertainty quantification and parameter optimization in simulations of the West African monsoon.
ISBN 978-3-7315-1442-8